The Need for Critical Thinking and the Scientific Method

The Need for Critical Thinking and the Scientific Method

Finlay MacRitchie

CRC Press
Taylor & Francis Group
Boca Raton London New York

CRC Press is an imprint of the
Taylor & Francis Group, an **informa** business

CRC Press
Taylor & Francis Group
6000 Broken Sound Parkway NW, Suite 300
Boca Raton, FL 33487-2742

Printed on acid-free paper

International Standard Book Number-13: 978-0-8153-6775-8 (Paperback)
International Standard Book Number-13: 978-0-8153-6815-1 (Hardback)

Library of Congress Cataloging-in-Publication Data

Names: MacRitchie, Finlay, author.
Title: Critical thinking and the scientific method / Finlay MacRitchie.
Description: Boca Raton : Taylor & Francis, 2018. | Includes bibliographical references.
Identifiers: LCCN 2017059910| ISBN 9780815367758 (pbk. : acid-free paper) | ISBN 9780815368151 (hardback :acid-free paper)
Subjects: LCSH: Science--Methodology. | Science--Social aspects. | Critical thinking.
Classification: LCC Q175 .M235 2018 | DDC 001.4/2--dc23
LC record available at https://lccn.loc.gov/2017059910

Visit the Taylor & Francis Web site at
http://www.taylorandfrancis.com

and the CRC Press Web site at
http://www.crcpress.com

Contents

Preface....xi
About the author....xv

Chapter 1 Introduction....1
Part 1: Common misunderstandings of the scientific method....1
The topic of debate must be defined unambiguously....1
A true scientist must be detached....2
A scientific theory can never be proven beyond doubt....2
Separate observations that are consistent with a theory do not correspond to additional confirmation....3
Refutability as a criterion for evaluating a scientific theory....3
Consensus is not a criterion for the validity of a theory....4
The illusion of modeling....4
Part 2: General introduction....4
The range of capacity for critical thinking....5
Distinction between science and pseudoscience....6
Transition from dogmatic to critical thinking....7
How this relates to anthropogenic global warming....8
Faults in application of the scientific method....8
Absence of scientific thinking in political debate....9
Summarizing thoughts....10

Chapter 2 The scientific method....13
The problem of demarcation....14
Notable theories of the early twentieth century....14
Comparisons of the different theories....15
Guidelines for evaluating theories....16
Refutability as the criterion for demarcation....16
Probability and informative content....17
Conjectures and refutations....18
Criteria for evaluating a theory....18
The need for science to grow (or to progress)....18

Requirements for progress of science.....19
Dangers to progress.....19
References.....20

Chapter 3 How the lack of scientific input impacts research organizations.....21
Australian ministers in charge of science portfolio.....21
The contribution of the Honorable R.G. Casey.....22
A short history of CSIRO after its formation.....23
Changes resulting from reviews of CSIRO.....23
How scientific knowledge is acquired.....24
How managerial control purports to drive science.....25
Comparison of the two approaches.....26
How has the CSIRO fared?.....27
The requirement to obtain a proportion of the research funding.....27
Effect of these changes on the workplace environment.....28
The climate change controversy.....29
The lack of scientific thinking in the debate.....29
Errors in application of science.....29
Balanced debate can still occur.....30
Expansion of the managerial approach.....30
The present working environment.....31
The Victims of CSIRO blog site.....31
References.....32

Chapter 4 How could this have happened?.....33
A career in science.....33
Effect of the requirement to procure funding.....34
Why do not more scientists enter politics?.....34
Are scientists different?.....35
How science of excellence can still be achieved.....36
How could it have happened?.....37
Making decisions without the basic knowledge.....37
How important scientific discoveries are made.....38
Effect of organizational changes on research.....38
Where angels fear to tread.....39
Where was the scientific leadership?.....39
Why do toxic work environments form in science organizations?.....40
How do toxic workplace environments arise?.....40
The control freak.....41
The narcissist.....41
The serial bully.....42
The psychopath.....42
What is a psychopath?.....43

Characteristics of psychopaths....43
How are psychopaths identified?....44
The Hare checklist....44
An example of psychopathic behavior....45
The legal argument....46
Psychopaths in science....46
What is the origin of psychopathic behavior?....47
Corporate psychopaths....47
How do psychopaths attain high positions in corporations?....48
Coping with psychopathic behavior....48
How it happened....49
Notes....50
References....50

Chapter 5 How the media influences public thinking....53
The myth of the lemmings....53
The invasion of Iraq in 2003....54
The Chilean military coup of 1973....55
How can we decide what is the truth?....56
Contrast between submissions to scientific journals and to the mainstream press....57
Issues that seem to be ignored by the mainstream media....58
Agenda 21 or sustainable development....58
British immigration since early 1980s....59
The opinion writers....60
How a democratic system can break down....62
The rise of social media....63
How to think for yourself....63
References....64

Chapter 6 Dangers to progress in science....65
Dangers to progress of science suggested by Popper....65
Obsession with number of publications....66
Quantity and quality in scientific publications....67
Citations and impact factors for measuring merit....68
Lack of citations to publications....69
An excess of poor-quality research....69
Negative flow-on effects from mediocre research....70
Not more scientists but more good scientists....71
Effects of external and internal forces....72
Misconduct in science....72
More subtle forms of unethical behavior....73
Cases....73
Conclusions from these cases....74

Fraud in science may be more prevalent than thought.....74
Misallocation of credit.....75
Maintaining the integrity of science.....75
Authoritarianism.....75
The earth as the center of the universe.....76
Lysenkoism.....76
Managerialism.....77
References.....78

Chapter 7 Applying scientific thinking to some current controversies.....79
Climate change.....79
Observations consistent with previous ones cannot be claimed as confirmations.....80
The weakness of modeling.....80
The medieval warm period.....81
The Great Barrier Reef.....82
Effects of acidification of the ocean.....83
The chemistry of carbon dioxide dissolution.....84
A critical examination is needed of the arguments for and against.....84
The saturated fat controversy.....85
Brief summary of the research on the effects of saturated fats.....85
Problems for food processors.....86
The problem of trans fats.....86
Replacement of trans fats.....87
The role of cholesterol.....87
LDL and HDL cholesterol.....87
A critical examination of the saturated fat–heart disease hypothesis.....88
Epidemiological research versus clinical studies.....88
Selection bias.....88
Encroachment of bureaucracy.....89
Status of the hypothesis.....89
The advent of gluten-free foods.....90
Celiac disease and gluten intolerance.....90
The need for gluten-free foods.....90
Explosion of gluten-free foods.....91
The downside of gluten-free diets.....91
Genetic engineering.....92
The possibilities for genetic manipulation.....92
Gene drives.....93
CRISPR.....93
Weighing the risks.....93

What are the risks?....94
A critical examination of the controversy....95
References....95

Chapter 8 Implementing scientific thinking and critical analysis......97
Research scientists in federal parliament....98
How to introduce more scientific thinking....100
Changes needed in the education system....100
The role of wonder in childhood education....101
Teaching with magic....102
Creativity....103
Introducing critical analysis....103
Some simple examples....104
Deflection of an issue....105
Putting labels on opponents....106
The straw man argument....106
Could have or might have....106
The prism of ideology....107
The influence of cultural background on the capacity for scientific thinking....107
References....107

Chapter 9 Bringing it together....109
Scientific/critical thinking....109
The scientific/critical versus the dogmatic approach....110
The diminished role of scientists....111
The effects on scientific excellence....112
How scientific organizations can be infiltrated by destroyers....113
The pros and cons of democracies....113
How can we determine if a proposed new law is better than the law it replaces?....114
It's the electorate, stupid!....114
The media....116
How public opinion can be influenced by the media....116
Opinion writers....117
Dangers to the progress of science....117
The lack of imagination or real interest....118
How managerialism erodes the standard of scientific research....118
Is science progressing?....118
If science is not advancing as it should, what is the cause?....119
How scientific progress can be eroded by authoritarianism....120
Flaws in the application of thinking to some current issues....120

Education as the means to raise the standard of thinking........123
Dangers posed by psychopaths........124
References........125

Chapter 10 Where will the future take us?........127
What do we learn from history?........127
What trends do we see?........128
How can these trends be reversed?........129
How can the standard of debate be improved?........129
Notes........130
References........130

Index........131

Preface

Democracy is the worst form of government except for all the rest according to Winston Churchill. But, for a democracy to be truly successful, the electorate needs to be made up of critical thinkers. If the people are not able to think critically, they are likely to swallow all the false information they are fed and democracy is put in danger of failing. Many present-day problems in the world are caused by people who are incapable of critical thinking and allow themselves to be manipulated and, in extreme cases, to be indoctrinated by false dogma. Critical thinking is a fundamental component of the scientific method. The philosophy of the scientific method is, unfortunately, not well understood and is even misunderstood by the general public. We regularly hear statements such as "the science is in," "it has been scientifically proven," or "there is scientific consensus that ..." The inferences from these statements are all false and, in some cases, are promulgated by people who call themselves scientists. Although the aim of science is to approach the truth, it does not ever claim to reach conclusions that are beyond doubt. Another common misconception is that when observations are made that are consistent with previous ones, each new observation counts as further confirmation of a theory, a type of reasoning that has been used by proponents of the theory of anthropogenic global warming. What this simply means is that each new observation can be interpreted in terms of consistency with the theory. It is easy to find confirmations if one looks for confirmations, but this is the dogmatic approach. The dogmatic approach does play a role in science in that it provides information on which to form hypotheses. However, it is based on induction, that is, making a series of observations and then attempting to arrive at a generalization from them. Induction is not a valid method for discovering knowledge, as has been argued by Karl Popper, one of the great philosophers of science. The true scientific method, the one that can advance genuine knowledge, is that of hypothesis-deduction. The critical approach that is used in the hypothetico-deductive method represents a transformation in thinking. Instead of trying to make observations in order to confirm a theory, the scientific approach designs experiments aimed at refuting or falsifying a theory. If the experiments result in a refutation, the theory is discarded and a new theory is sought. The new theory makes use of what has been learned from the refuted theory. In that way, science proceeds by a trial-and-error procedure. If, on the other hand, the new experiments fail to refute the theory, then the theory is corroborated but remains open to possible refutation by further critical experiments or rational debate.

The scientific approach thus applies critical thinking to tackle problems, and true scientists are prepared to listen to arguments from all sides and to change their opinions if the arguments are logically convincing. It might be expected that those with such attributes would be valuable for influencing decision making in society, yet there is a conspicuous absence of scientists in political life. Even worse, decisions about scientific matters are not being made with input from scientists but are often made by those from other professions, notably law, economics, and business management. One of the consequences has been the removal of top scientists from directing scientific organizations and their replacement by those with management qualifications. A glaring example is what has happened to the Australian government's premier research organization, the Commonwealth Scientific and Industrial Research Organisation (CSIRO). The supplanting of scientific leadership by managerialism in the past few decades has seen the erosion of scientific excellence and a loss of morale among its scientists. The encroachment of a managerial culture into a scientific organization with imposition of an alien mode of thinking is responsible for these negative outcomes. The inappropriateness of this cultural shift has opened the way for impostors to enter the hierarchical structure. These individuals wreak havoc on an organization just as they have been responsible for the failures in leadership that have caused the destruction of previously successful corporations that we have been witnessing in recent times.

Science is unique in the sense that, by its nature, errors are systematically criticized and, in time, corrected. This leads to progress, whereas, in most other human endeavors, there is change but not necessarily progress. There are, unfortunately, forces at work that are pressing to destroy this uniqueness. Scientific research should be a relentless pursuit of truth. However, in the performance evaluation of scientists, the number of published papers is being adopted as a useful criterion. This is tempting some scientists to perform superficial studies to try to maximize their number of publications rather than carrying out in-depth investigations. There has been a concomitant explosion in the number of scientific journals and, of course, the number of published papers. This, together with the pressure to publish, causes some researchers to fail to carry out due diligence on the existing literature, resulting in important previous work being overlooked or not given appropriate credit. It can result in the reinvention of the wheel, but the new wheel often does not work as well as the old one.

We are presently seeing a deficiency of critical thinking everywhere. Conflicts around the world are fanned by political manipulation and indoctrination, often with support from the media. Political partisanship runs rife to such an extent that the primary objective appears to be to attack the opposing parties by distorting the truth and abusing adversaries, rather than doing what is best for the national interest. How can all this be changed so as to develop more bipartisanship and cooperation for improving humankind? The impetus for change has to come from the general public. Politicians and leaders will not change unless the public demands it. If what they are doing is working, there is no incentive to change. If you can twist the issues to convince a public that is largely ignorant and gullible, why change the strategy? The only path to change is to educate the public so that they will think critically and insist on getting the truth. This is not easy to do, but we must try. Cultures that base their actions on hate for other cultures, which may have endured for generations, present a formidable challenge. Teaching of critical thinking should begin at an early age, preferably at primary school. There should also be instruction on what constitutes the scientific method so that there will not be the misunderstanding of science that exists in a large proportion of the adult population.

About the author

Finlay MacRitchie was a professor in the Department of Grain Science & Industry, Kansas State University from 1997 to 2009. He is presently professor emeritus in that department. Prior to this, he was a research scientist in the Commonwealth Scientific and Industrial Research Organisation (CSIRO) of Australia. He has spent short periods of time as visiting professor at the University of Chile and the Federal University of Rio de Janeiro, Brazil, and as senior research fellow at the Agricultural University of Wageningen, The Netherlands; the University of Paris V; the University of Lund, Sweden; and the University of Tuscia, Viterbo, Italy.

He has published more than 150 papers in refereed journals and three textbooks: *Chemistry at Interfaces* (Academic Press, 1990), *Concepts in Cereal Chemistry* (Taylor & Francis, 2010), and *Scientific Research as a Career* (Taylor & Francis, 2011).

MacRitchie's awards include the F.B Guthrie Medal of the Cereal Division of the Royal Australian Chemical Institute, and the Thomas Burr Osborne Medal and George W. Scott Blair Memorial Award of the American Association of Cereal Chemists (now AACC International).

He has been a member of the editorial boards of *Advances in Colloid and Interface Science*, *Cereal Chemistry*, and *Journal of Cereal Science* and editor-in-chief of the *Journal of Cereal Science*.

chapter one

Introduction

Part 1: Common misunderstandings of the scientific method

Scientific thinking embodies two main concepts. One is an understanding of the scientific method and the other is what is usually referred to as critical thinking. There is an acute lack of understanding of the scientific method among the population and this is leading to serious deficiencies in the public debate. To illustrate this, we are going to scrutinize just one example of the issues that are currently controversial and which involve science. The example chosen is that of anthropogenic global warming or which is more loosely described as climate change. Immediately, we come up against a problem.

The topic of debate must be defined unambiguously

Most rational people accept that the climate has changed in the past, is currently changing, and will likely continue to change in the future. They also realize that the topic being debated is that of anthropogenic global warming; that is, that there is a significant (measureable) contribution to warming of the planet by human activities. The main human activity is the burning of fossil fuels, which produces what are called "greenhouse gases," of which carbon dioxide is one. These gases then concentrate in the upper atmosphere and act as a sort of blanket that hinders the escape of heat. The scientific basis of this effect is generally accepted. What is not known with any certainty is the magnitude of the effect. The sloppiness in defining the issue can be a source of confusion. Furthermore, other confusing terms are bandied about such as "carbon pollution" and "putting a price on carbon."

Other current topics of debate suffer from the same lack of clarity in assigning names. Examples are "same-sex marriage," which has morphed into "marriage equality," and another is "multiculturalism," which can cover a wide range of activities. In some cases, the title of the topic is deliberately manipulated so as to cause obfuscation.

In summary, in order to have a valid scientific debate, it is essential that the topic of debate is unambiguous to ensure that the participants are arguing about exactly the same thing.

A true scientist must be detached

The most crucial change in the evolution of science has been the transition from dogmatic thinking (pseudoscience) to critical thinking (true science). In simple terms, dogmatic thinking involves looking for confirmations of a theory whereas critical thinking is open to observations that do not agree with a given theory. A scientist who uses critical thinking forms opinions, but these are not held rigidly and can be changed or modified by new observations or rational arguments.

What do we find in the debate on anthropogenic global warming? The participants are mostly divided into two groups: those who believe it is occurring and those who do not. This has led to the terms "alarmists" and "deniers." Placing labels on those who do not agree with your viewpoint is immature and is not in the spirit of true science. True scientists should be detached so as to be able to listen carefully to contrary arguments and be ready to change their opinions. Many of the protagonists in the debate are fierce in asserting their beliefs and casting scorn on their opponents. This is not how an issue should be scientifically debated.

To hold a reasoned scientific debate requires the participants to be sufficiently detached to be open to opposing arguments and to change or, at least, modify their opinion should the argument be logically strong. In cases where the evidence is inconclusive, it is best to defer making a judgment until more light is thrown on the issue.

A scientific theory can never be proven beyond doubt

In the debate about anthropogenic global warming, we hear statements such as "the science has been settled." Those who make such statements do not understand science. The truth is that the science is never settled, especially for such a complex issue as the one we are considering. The best we can hope for is that research will increase our understanding of the problem and bring us closer to the truth. True scientists retain a certain skepticism so that they always remain open to new perspectives. Dogmatic thinkers close their minds to any ideas that may threaten their established views.

If we examine the approaches being used to throw light on the question of anthropogenic global warming, we see that practically all of the effort has been aimed at trying to find confirmations, either for or against. Have any attempts been made to refute the theory? I am not aware of any. In fact, if only dogmatic thinking is applied to an issue, the idea of trying to falsify one's strongly held belief is unthinkable.

Separate observations that are consistent with a theory do not correspond to additional confirmation

Many believe that, if an observation is consistent with a theory, then further separate observations that are also consistent, signify additional confirmation. Perhaps an example may help to illustrate what is a subtle point. In the debate on anthropogenic global warming, it has been observed that in some parts of the world glaciers are melting. A further observation is that sea levels in certain parts appear to be rising, a result that would be expected if the temperature were increasing. This then is taken as additional confirmation of the theory that the planet is warming. This assumption has, however, been argued by Karl Popper, one of the great philosophers of science, to be not valid.

What does this additional observation confirm? It simply means that each observation can be interpreted in the light of the theory. If we only look for confirmations, we are bound to find them. If we look for confirmations that the planet is not warming, we can also find them. For example, the polar ice cap in a certain area of Antarctica is increasing in size. The temperature of the earth's atmosphere has also not been found to have increased over a recent period of about eighteen years, even though the amount of carbon dioxide being emitted from industrial processes has been steadily increasing.

All these observations are important to throw light on the issue and for scientists to form hypotheses. However, they do not necessarily bring us closer to an acceptable theory.

Refutability as a criterion for evaluating a scientific theory

How then, we may ask, is it possible to test a theory so as to give us an answer that can bring us closer to the truth and provides us with more confidence that we are on the right track? Karl Popper wrestled with this problem and came up with a criterion. Instead of searching for confirmations of a theory, we should try to show that it is false. This approach represented the evolution from dogmatic thinking to critical thinking. It opened a way to critically test a theory by trying to refute it. If the experiment succeeds in a refutation, then the theory is discarded and a search for a new theory is commenced. If an experiment to severely test the theory fails to refute it, then we can say that the theory is corroborated for the present but with the proviso that it may be refuted by further experiments in the future.

Consensus is not a criterion for the validity of a theory

Claims are often presented for the validity of a theory by pointing to the number of scientists who support it. The implicit argument is that if 90 percent of scientists support a theory, then this must mean that it is correct. Perhaps 99.9 percent of scientists believe a theory to be true. Therefore, it is inferred that there is an even greater certainty that it must be true. Such logic (or lack thereof) cannot be justified. Many of the great discoveries of science have been made by individuals who have not followed consensus. In fact, it is due to this that they have pioneered new ways of thinking. It is probably a requisite for new advances that they separate themselves from the mainstream.

The illusion of modeling

In recent times, modeling has been developed to try to describe events. It has been spurred on by the advances in the power of computing. The aim in modeling is to input all the variables that are believed to influence a given phenomenon. This then can produce an output that gives an answer to a certain question. For this approach to work requires that all the variables that affect an outcome are fed in and that they are fed in a correct manner. In a way, this approach applies the idea proposed by Popper of devising experiments to refute a theory. If the data are correctly introduced, the result predicted by the theory can be either corroborated or refuted. Providing the problem being tackled is fairly simple, this approach can be successful.

Modeling has been applied to try to describe the effects of human activity on climate. The way that climate changes is a complex problem and the computer programs that have been applied are huge. So far, they have not been able to accurately predict the effects that are being observed. How to interpret this is equivocal. It may be that the theory is not correct so that would amount to a refutation. On the other hand, it may be that the input of information is flawed so that no firm conclusion is justified.

Part 2: General introduction

The way in which individuals form opinions about issues in our society is influenced in two main ways. First, we tend to read books and articles or listen to viewpoints that are in accord with the opinions we have formed from our experiences, thus reinforcing bias. Second, much of the public debate is characterized by distortions of the truth, albeit usually subtle, and, unless alert to this, our thinking may be prejudiced. What is needed to put aside our bias and be able to recognize the lack of truth in what we hear or read? How effectively we deal with this question depends on our capacity for critical thought. Critical thinking is an essential ingredient of the scientific method. This does not mean that we who call ourselves scientists are the only ones who can provide the answers. Scientists are human and suffer from the same failings as nonscientists. We can be gullible, opinionated, and intolerant of the views of those who hold different opinions. What I would like to suggest is that there is a true science and the practitioners of true science are those who strive to uphold its ideals. To do this does not require formal qualifications in science but requires an understanding of what is inherent in the scientific method and how it should be applied. It is therefore an idealized concept perhaps analogous to that of the philosophers portrayed in Plato's *Republic*. True scientists are ones who follow the scientific method, which is outlined in Chapter 2. A true scientist brings an unbiased attitude to debate, does not adopt dogmatic viewpoints, is prepared to respect the opinions of those with opposing viewpoints, and is ready to modify his or her opinion if sound counter arguments are put forward.

Sometimes, in this book, reference is made to scientific thinking. This term is meant to embody critical thinking combined with an understanding of the scientific method.

The range of capacity for critical thinking

The capacity for critical thinking is what can determine the contribution that each individual makes to the debate of an issue. We can think of two extremes. There are the critical thinkers who free themselves from the constraints of bias and the distortions of the media. At the opposite extreme, we have those who allow themselves to be indoctrinated and finish up with a narrow dogmatic view of the world. If this simple fact could be recognized by those with an uncritical frame of mind, they would see that they are mere pawns who are exploited and have little control of their destinies. This can be shown by a simple thought experiment. Suppose we have two different cultures in which there is mutual hate and intolerance between them. If we clandestinely take a baby from one culture and exchange it with a baby from the other culture, what will be the likely outcome? The most likely outcome is that each baby will become an adult and adopt the beliefs and hates of the culture into which they have been planted. In other words, they are simply puppets who, although entering the world as humans with all the potential to form high ideals, develop characteristics determined as a result of indoctrination by the society in which they are placed.

Much of the evil that occurs in the world is caused by indoctrination with lies. The puppets who absorb the indoctrination are individuals with little powers of critical thought. Of course, all of us are susceptible to some degree of indoctrination. Perhaps if we could spend more time contemplating such things as a star-studded sky on a clear moonless night far from city lights, witness waves crashing onto a deserted beach, or take in the majesty of giant trees in a forest, this might instill a sense of wonder that could help to break the shackles that bind us to the petty hates and intolerances that tend to envelop us. The main theme in this book is that we can all gain by acquiring greater understanding of the true scientific method and how it can be applied to our thinking.

Distinction between science and pseudoscience

Science involves a search for truth. The philosophy of the scientific method is discussed in detail in Chapter 2. A belief held by many is that science is characterized by its empirical nature; that is, it is based on observations and measurements. To some extent, this is true but it is not the complete picture. For example, astrology is based on observations and measurements of the positions of heavenly bodies but is not regarded as a science but more as a pseudoscience. The distinction between science and non-science or pseudoscience is elaborated in Chapter 2. Put in its most simple form, it is that a valid scientific hypothesis is testable and therefore, in principle, refutable, whereas a nonscientific hypothesis is not. We may briefly illustrate this by considering the issue previously mentioned to use as an example, that of anthropogenic global warming. It is doubtful if any rational person would not agree that the earth's climate may be changing, has been changing in the past, and likely will continue to change in the future. Of course, nearly everyone understands that the current debate is about whether there is a significant anthropogenic (caused by humans) contribution to warming of the planet and that the effect is due to emission of greenhouse gases, of which carbon dioxide is one. Even so, in a scientific debate, it is always imperative to strictly define an issue so that there can be no ambiguity and to ensure that people are arguing about the same thing.

There are many reported observations that are consistent with the belief that the earth's atmosphere is warming and that this is due to the greater emissions of greenhouse gases resulting from increased industrial activity. The observations that have been reported include declining arctic sea ice, retracting glaciers, and sea level rises resulting from warming of the oceans. Observations that have been put forward to present a contrary view are that there has been no global temperature rise during a recent period of some eighteen or so years, despite an ever increasing emission of greenhouse gases, and that historical data suggests that a cyclical ice age is imminent. There is a considerable amount of information available in the literature dealing with these different observations. This controversy is deliberately not discussed here in order to avoid any accusation of selective referencing. The intention is not to analyze the information to arrive at an opinion one way or the other. It is to use the issue to try to illustrate that there has been many shortcomings in how the debate has proceeded and how true scientific thinking should be applied.

Transition from dogmatic to critical thinking

In the evolution of the scientific method, the starting point involves myths. Myths are characterized by dogmatic thinking, that is, attempts to confirm theories by selecting those observations that are consistent with the theory or hypothesis. The scientific method represents a fundamental transition to critical thinking in which assumptions are questioned and experiments are designed to severely test a theory and to try to refute it. Should a refutation not be obtained, the theory is considered to be corroborated. This means that it is tentatively held to be consistent with the facts although it can never be proven to be true beyond doubt. In contrast, if a severe experimental test refutes a theory, the theory must be rejected and a search for a new theory undertaken. The new theory can make use of what has been learned from the refuted one. Scientific research thus progresses by a trial-and-error procedure in which we learn from our mistakes. In simple terms, a scientific theory can never be unquestionably proven to be right but it can be proved to be wrong. We hear statements such as "the science of global warming is settled." The implication is that the issue is proven. The truth, as we have seen, is that the science can never be "settled."

The path to approaching the truth using the scientific method thus involves testing hypotheses, rejecting those that are shown to be false and proceeding to form new hypotheses. In that way, we get closer to the absolute truth, although we can never assume that we can attain it. The terms *theory* and *hypothesis* will be used interchangeably throughout the book. However, it is usual to distinguish between them. A hypothesis refers to a conjecture that has not been subjected to much testing, whereas a theory is considered to have undergone a fair amount of testing and has prevailed in the face of serious attempts to refute it.

We therefore see a crucial distinction between the early stages of scientific evolution, characterized by dogmatic thinking and the more advanced stages that involve critical thinking. Dogmatic thinking implies attempts to find confirmations. It is relatively easy to find confirmations of a theory if we look for confirmations. When this is the goal, it is more likely that we will ignore or downplay observations that do not confirm the theory. In contrast, the critical approach actively engages in a search to find exceptions to the theory. The dogmatic approach looks for confirmations; the critical approach looks for refutations or falsifications.

How this relates to anthropogenic global warming

How do these ideas relate to the issue that has been selected as an example of a scientific debate, that of anthropogenic global warming? It is hard to find evidence of critical thinking in this debate. Most who engage in the debate have formed an opinion that they steadfastly adhere to. Have any experiments been designed to try to refute the theory, one way or the other? One of the ploys used is for those who hold a particular view to attach labels to their opponents (i.e., those with the opposing view). Thus, those who support the validity of anthropogenic global warming are branded as "alarmists," whereas those who disagree are called "deniers." The aim is to denigrate those with an opposite viewpoint so as to suggest that their opinions should not be considered seriously. Many of those who participate have scientific qualifications and call themselves scientists. The contrast with true science could not be more stark. The scientific thinking discussed in this book is meant to refer to an idealized way of thinking. This involves approaching an issue with an open mind devoid of bias, holding an opinion based on the evidence available, but willing to listen attentively to those who disagree and be disposed to change the opinion in response to new experimental results or rational argument. Unfortunately, we scientists are human and do not always uphold these virtues. We can be opinionated, dogmatic, and intolerant of those with opposite viewpoints, the same as non-scientists.

Examination of the arguments that have been made in this debate shows that they are almost entirely based on dogmatic thinking. They are not based on true science but on nonscience or pseudoscience. When we look for observations that appear to agree with our theory, each successive observation is claimed to be a further confirmation. This is not a valid assumption. All it means is that each successive observation can be interpreted in terms of consistency with the theory. The only observations that count are ones that result from making risky predictions aimed at falsifying the theory, as is explained in more detail in Chapter 2. This is the critical approach that is inherent in the true scientific method. The challenge for those who genuinely wish to approach the truth is to come up with experiments that can test the theory by inviting refutations. So far, this does not seem to have been done to any extent. Of course, this does not necessarily mean that new information always needs to be acquired. It is possible to propose a hypothesis and test it by predicting what should have happened in the past.

Faults in application of the scientific method

It is useful, at this stage, to stress some other misunderstandings of the scientific method that have been used in arguments and which have been briefly noted at the beginning of the chapter. One is the belief that the opinion of the majority of scientists should be accepted. In science, consensus counts for nothing. In fact, most of the great scientific discoveries have been made by those who have not followed consensus. Another concept that has been given unwarranted importance in the global warming debate as well as in other debates has been that of modeling. In modeling, a model is first created. There is an input in which all the variables considered to be relevant to the problem being tackled are fed in. There is then a corresponding output that gives an answer to the question being asked. This approach can work well if the question being pursued is relatively simple. Modeling can give a reliable answer if all the variables that impact the result are known and if they are introduced into the model correctly. In the case of global warming due to human emissions, this is an extremely complex problem. It is doubtful if all the relevant variables are understood so that the input may very well depend on untested assumptions. Garbage in, garbage out is an expression that has been coined to describe such an exercise.

Many of the current debates on issues involving scientific aspects suffer from the same deficiencies as those that are evident in the debate on anthropogenic global warming. Decisions resulting from these deficiencies can lead to far-reaching effects on the culture and economic development of countries. For example, the acceptance that greenhouse gas emissions are making a significant contribution to warming of the planet has led to imposts on companies that produce these emissions. These imposts are then passed on to consumers. The net effect is to reduce the economic wealth of the country and drive the population toward poverty rather than affluence. The justification for curbing emissions is that possible consequences of global warming may also incur negative effects on economies by such events as crop failures. Furthermore, reduction of pollutants in the atmosphere leads to improvement of the environment and, as a result, human health. It could eventually turn out that science will show that human emissions do cause significant warming of the planet. However, the point is that this does not seem to have happened so far. Thus, all the decisions that have been made to counter potential global warming have not been based on true science, although this is what has been claimed by its proponents.

Absence of scientific thinking in political debate

True scientific thinking does not have much influence in political decision making. Very few scientists attain positions of power in public life, largely because they do not seek to. Most of the decision making falls to other professions, for example, lawyers, economists, and business managers. The other groups of influential people are those in the media, many of whom are not individuals with a strong understanding of science. Law is considered to be one of the most appropriate stepping-stones for a career in politics. Law differs from science in several fundamental ways. One is the way it uses adversarial principles. Science can also be adversarial and scientists frequently participate in vigorous debate. The difference is that lawyers are usually unflinching in their adherence to an argument as they must be if defending someone charged with a crime. True scientists, in contrast, are prepared to change their viewpoint as a result of new evidence or sound reasoning. Another difference is that, whereas in law, decisions are often based on precedence, in science, precedence counts for nothing. Each and every interpretation or theory is continually questioned and challenged, regardless of the time it was originally proposed. Scientific understanding is a dynamic process and always subject to alteration.

When statements that are made every day by politicians and reporters are critically examined, it is relatively easy for discriminating people to detect the spin, bureauspeak, and empty rhetoric that often permeate the glib language. Public debate often involves defending an entrenched opinion rather than equitably considering opposite points of view. Unfortunately, it seems that a large proportion of the public do not bring an open mind to issues and accept what they want to hear rather than critically evaluate an argument. Many people hold dogmatic opinions and are not disposed to change them. True scientific thinking, in contrast, involves entering a debate without preconceived opinions and considering the various arguments impartially.

Many of those who participate in a debate of a scientific nature are influenced by vested interests. For example, departments of climate change have been created, many with the mission to minimize the supposed future effects. Those appointed to positions in these departments naturally are expected to toe the line. Their approach, as a result, is far from impartial. Similarly, many researchers who work on the issue of climate change depend on financial support. In order to guarantee continuing funding, they need to present conclusions that agree with the aims of those who control the purse strings. Their research will therefore concentrate on seeking results that agree with those aims and not on designing experiments to try to falsify them. In other words, based on the philosophy of the scientific method so far discussed, their activities cannot be described as scientific but fall into the realm of pseudoscience.

Summarizing thoughts

If an issue is to be debated scientifically, it is imperative that it be defined clearly and unambiguously. This is not always the case in public debate as we have seen with the "climate change" debate and many others. For example, the issue of "multiculturalism" is often vigorously argued.

For some, it is accepted to mean that immigrants integrate into the society of their new country while bringing with them customs that enrich the culture. Others think of it as formation of enclaves of different cultures having little interest in adopting the culture of their new country. Between these two extremes, there exist varying degrees of harmony. No effective debate can result from this ambiguity.

When many different observations are made that are consistent with a theory, it is often concluded that each successive observation counts as a further confirmation of the theory. This assumption is not valid. It only means that the observation can be interpreted in terms of consistency with the theory. The only experiments that count are those that are designed to falsify (refute) the theory. If the experiment fails to do this, then the theory is corroborated and may be tentatively held to be true but it can never be proven. The change from seeking confirmations of a theory to seeking refutations represents the change from dogmatic thinking to critical thinking and thus to true scientific thinking.

How then should we try to encourage true scientific thinking? We have to accept that it will not be some quick fix. It requires a great deal of patience and resolve to make even a small impact. First, we should denounce statements that are ambiguous and explain why they are unacceptable. Then, we need to vehemently denounce statements such as "it has been scientifically proven" or "the science is settled" and explain why they are false. When we debate issues, we need to try to free ourselves from dogmatic opinions, be prepared to listen sympathetically to opposing views, and be ready to admit that they have merit. Does this mean that we should adopt this approach with everyone with whom we have discussions? Certainly not! It is only possible to have a worthwhile debate with others who are also prepared to go at least some of the way toward returning your good will. There is no point in casting pearls before swine. There are those who it is quite useless to argue against. Some have such ingrained opinions that they would opt to kill you rather than entering into a discussion in which the beliefs that they dogmatically hold might be challenged. Unfortunately, the inability to consider the views of others is what inevitably leads to much of the conflict and violence that permeates the human world.

Can the status quo be changed and, if so, how? The obvious way is through education. Is critical thinking included as an important subject in school curricula? It does not seem to be. In fact, the opposite appears to be true. Those who plan and control school curricula, as well as the teachers who implement them, tend to predominantly hold particular political views and try to instill their ideology on to students. In order to counter this, the subject of critical thinking needs to be introduced into school curricula, preferably at an early level in primary school by competent instructors. In addition, the true philosophy of the scientific method needs to be explained and discussed.

chapter two

The scientific method

What is the scientific method and how does it differ from other methods of inquiry? As noted in Chapter 1, science is empirical but this cannot be its only distinguishing feature. Science has evolved over the centuries. Its beginnings initially involved observations made by inquiring minds. With the passage of time, observations were subjected to measurement and methods were invented for doing this accurately. Then, apparently unconnected observations were tied together by generalizations or theories. This same process occurs during the evolution of a particular theory. The early stages of development of a theory tend to look for confirmations of the theory. This essentially is the inductive approach. In simple terms, induction involves making a series of observations and then attempting to arrive at a generalization from them. One of the great philosophers of science, Karl Popper, in his book *Conjectures and Refutations* (Popper, 2002), has shown that induction is not a valid method for advancing knowledge. Popper's analysis of induction followed that of Hume, who pointed out that induction cannot be logically justified. The basis of Hume's analysis was that the theory of induction leads to an infinite regress based on repetitions. Each interpretation of an observation is based on a previous interpretation of an observation. Popper proposed that the theory of induction be replaced by a different concept. Instead of passively waiting for repetitions to impose regularities on us, we try to impose regularities on the world in the form of hypotheses. In the hypothetico-deductive method, an inspired guess (conjecture) or hypothesis is made to attempt to explain observations. Experiments are then designed to test the hypothesis. Rather than trying to obtain a confirmation, these experiments aim to severely test it. There is thus an essential difference between the two methods. Whereas in the inductive approach, scientific knowledge was claimed to proceed from observation to theory, in the hypothetico-deductive method it proceeded from theory to observation. Instead of waiting for regularities to appear, scientists actively try to impose regularities on nature. This then leads to critical tests that are not arbitrary nor accidental but undertaken with the intention of testing the validity of the hypothesis. Frequently, nature resists and the hypothesis is refuted. A refutation should not be considered as a failure. It means the elimination of one possible explanation and stimulates further thinking aimed at creating a new hypothesis.

It should be noted that there have been many others besides Popper who have made important contributions to the philosophy of science. For example, Thomas Kuhn (1962), in his book *The Structure of Scientific Revolutions*, suggested that science progresses not through a linear accumulation of new knowledge but by periodic revolutions in which there are abrupt transformations, later referred to as paradigm shifts. Other eminent philosophers of science who have contributed books include Stephen Toulmin (1953), Carl Kordig (1975), Paul Feyerabend (1975), and Norwood Hanson (1958). As a result of my experience in scientific research over many decades, I have come to feel much closer to Karl Popper than other philosophers of science, so that it will be Popper's thinking and concepts that I have chosen to be the basis for discussion in the remainder of the book.

The problem of demarcation

Hypotheses cannot be created from nothing. They have to be preceded by observations and attempts to generalize them. Thus, induction plays a part in science. It provides the information needed to form hypotheses. It can thus be considered as an earlier stage in the evolution of a scientific theory. The proposition introduced by Popper was a crucial turning point in the understanding of the scientific method. It is instructive to examine how he arrived at this point. In the early part of the twentieth century, Popper wrestled with what he called the problem of "demarcation." This was the problem of how to distinguish between science and nonscience or pseudoscience. He saw the problem not as to decide if a theory or hypothesis was correct but to search for a criterion to determine if it was scientific or not. For example, what is the essential feature that prevents us from accepting astrology as a scientific discipline?

Notable theories of the early twentieth century

In the intellectual climate in which Popper grew up in Vienna, Austria, there were several important theories being actively discussed. These included Adler's theory of individual psychology, Freud's psychoanalysis, Marx's theory of history, and Einstein's theory of relativity. Freud believed that personality was shaped by experiences in early life that were incorporated in the unconscious component of behavior and which later could be manifested in the conscious mind. In the other psychological theory, Adler saw people as being motivated mostly by social influences and their striving for superiority or success. Marx viewed history as a conflict between an ownership class that controlled and a proletariat class that provided labor for production. History was seen to progress through a class struggle in which capitalism would eventually self-destruct and be replaced by a stateless, classless society. Einstein's theory of relativity introduced new concepts into understanding the physics of the universe. When problems involving large distances and high velocities such as that of light travel are considered, it was necessary to include the parameter of time and to replace 3-D space by a 4-D space–time continuum.

Comparisons of the different theories

What struck Popper about the psychological theories of Adler and Freud was that their adherents were always finding verifications. In fact, there seemed to be no instances of behavior that could not be explained in terms of the theories. This caused Popper to suspect that, far from being a strength, as their proponents claimed, this could be their weakness. On one occasion when he was accompanying Adler in his social work with the youth of Vienna, Popper related an example of behavior that he had difficulty in understanding. Adler, however, immediately explained it on the basis of his theory. When asked how he had reached that conclusion, Adler replied that it was because of his thousand-fold experiences. Popper realized that each of the thousand observations, rather than counting as additional confirmations, meant nothing more than that each could be interpreted in terms of being consistent with the theory.

Einstein's theory was quite different. One of the predictions of the theory of relativity was that light should be deflected by gravitation. In 1919, this prediction was put to the test by measuring the positions of distant stars when their light passed near a heavy body, the sun, during a total eclipse compared to their positions in the night sky. The theory of relativity not only correctly predicted that the light would be deflected but also the direction and magnitude of the deflection. Popper concluded that the criterion for a scientific theory was that it must be refutable. The theory of relativity fitted this criterion. Even if the light had been deflected but the magnitude of its deflection had been different to what had been predicted, the theory would have been refuted. In contrast, although Adler and Freud may have been seeing many things correctly, their theories could not be considered to be scientific because it was not possible to conceive an experiment that could refute them. In the case of the theory of Marx, the theory was refutable, but Popper asserted that it had been tested and was refuted. However, in order to preserve it, its adherents had introduced auxiliary hypotheses and reinterpreted the theory and the evidence to make them agree. This is possible to do but can only be achieved by a loss in scientific status.

Guidelines for evaluating theories

Popper's deliberations led him to formulate the following conclusions (Popper, 2002):

1. It is easy to obtain conformations or verifications for nearly every theory—if we look for confirmations.
2. Confirmations should only count if they are the result of risky predictions: that is to say, if unenlightened by the theory in question, we should have expected an event which was incompatible with the theory—an event which would have refuted the theory.
3. Every "good" scientific theory is a prohibition: it forbids certain things to happen. The more a theory forbids, the better it is.
4. A theory which is not refutable by any conceivable event is non-scientific. Irrefutability is not a virtue (as people often think) but a vice.
5. Every genuine test of a theory is an attempt to falsify it, or to refute it. Testability is falsifiability; but there are degrees of testability: some theories are more testable, more exposed to refutation, than others; they take, as it were, greater risks.
6. Confirming evidence should not count except when it is the result of a genuine test of the theory; and this means that it can be presented as a serious but unsuccessful attempt to falsify the theory (I now speak in such cases as corroborating evidence).

Refutability as the criterion for demarcation

Thus, Popper's search for a criterion for demarcation (that which distinguishes science from pseudoscience) led him to a very simple concept. A theory is scientific if it is refutable (or testable or falsifiable). In the case of the theories of Adler and Freud, every observation of behavior could be explained by the theories. No experiment could be devised that could refute them; they were irrefutable. This property of irrefutability was held up by their proponents as a confirmation of their validity. In Popper's view, irrefutability meant just the opposite. It eliminated them from being considered as scientific.

Although refutability is a relatively simple concept, there are degrees of refutability and therefore there are degrees in the merit of a theory. Popper argued that the refutability of a theory is related to its *a priori* probability. The lower the probability of a theory, the better it is scientifically. This may appear, at first thought, to be counterintuitive, as many believe that a high probability should be synonymous with a good theory. To illustrate the concept in a rather simplistic (and nonmathematical) way, let us compare a prediction from the theory of Adler with one from the theory of Einstein mentioned earlier. Instances of behavior in terms of Adler's theory were always highly predictable, hence highly probable. In a sense, they were similar to the predictions of fortune-tellers. Fortune-tellers predict events that are so general that they have a high probability of occurring. They may say that something good will happen to you tomorrow. If you happen to wake up in the morning, that could be taken as a good thing. It is almost certain to happen so it has a high probability. In contrast, the theory of Einstein predicted something that no one would have expected on *a priori* grounds, that is, that light would be deflected by a heavy body and the amount of deflection could be quantitatively estimated.

Probability and informative content

If the predictions of a theory are absolutely probable, it means that the theory goes no (or very little) further than present knowledge. The theories that are most highly regarded in science are those that depend on risky predictions (e.g., that light should be deflected by heavy bodies according to Einstein's relativity theory) and therefore have a low probability. Theories that have a low probability can be shown to have a high informative content. The consequence is that they have the potential to expand the boundaries of knowledge beyond what was initially sought. Progress in science, on this line of thinking, means progress toward more interesting, less trivial, and therefore less probable theories. Popper used the word "verisimilitude" (truthlikeness) to describe the property that is sought in good scientific theories. Verisimilitude combines truth and content, whereas probability combines truth with lack of content.

Popper illustrated the relation between probability and content by a simple example. If a is the statement "It will rain on Friday" and b is the statement "It will be fine on Saturday," then ab is the statement "It will rain on Friday and it will be fine on Saturday." The information (content) of the conjunction ab thus exceeds that of its component a and its component b separately. It follows that the probability that ab is true will be smaller than that of either a or b separately. Mathematically, if we write Ct for the content of a statement and p for the probability, then

$$Ct(a) \leq Ct(ab) \geq Ct(b)$$

and

$$p(a) \geq p(ab) \leq p(b)$$

Conjectures and refutations

The title of Karl Popper's book is a fitting one to describe the way the scientific method proceeds. It may begin with an inspired guess to try to relate previously unconnected observations. This guess (hypothesis, conjecture) is then tested by experiment and/or rational criticism. The better the conjecture, the more testable it will be. If the experimental test or rational criticism shows the hypothesis to be invalid, then this is accepted and a search is commenced for an alternative hypothesis. This search will be helped by ideas embodied in the failed hypothesis and by the experiments/arguments that were used to refute it. If the initial experimental tests and criticisms do not succeed in refuting the hypothesis, then it is held to be corroborated. Subsequent thinking may cause the hypothesis to be modified and improved. Thus, the scientific method is a trial-and-error procedure in which hypotheses (theories) are tested, tentatively held, or discarded, and new hypotheses proposed. It is important to realize that a refuted hypothesis may play an important role in the development of a new hypothesis and is a vital part of the trial-and-error scientific method. Refutations are often mistakenly regarded as establishing the failure of scientists or their theories. The truth is that every refutation should be considered as a success, not only a success of the scientist who refuted the hypothesis but also of the scientist who created the refuted hypothesis and who, perhaps indirectly, suggested the refuting experiment.

Criteria for evaluating a theory

Popper's deliberations led to the concept of relative potential satisfactoriness of a theory. Even before a theory is subjected to testing, its worth as a theory can be assessed. One theory is better than another if it can lead to a greater amount of empirical information or content as well as being logically superior. It is better than another if it has a greater explanatory and predictive power and can therefore be more severely tested by comparing predicted facts with observations. Put succinctly, a better theory is one that has a higher degree of empirical content or testability.

The need for science to grow (or to progress)

Popper emphasizes an important aspect of science: its need to grow. The growth of scientific knowledge does not mean a continuous accumulation of observations but rather a critical examination of theories and their replacement by better ones. This progression can be clearly seen in the cosmological theories from Kepler and Galileo to Newton and Einstein. Kepler and Galileo were great scientists who developed theories that those who followed (Newton, Einstein) could build on and produce better theories. Critical examination of theories leads to attempts to overthrow them. This then results in further experiments and observations of a kind that nobody would have imagined without the stimulus of the earlier theories and criticisms of them. In this way, scientists approach more closely to the truth without ever knowing how close they have come to it. The greatest contribution that a theory can make to the growth of scientific knowledge can be the new problems it raises. The growth of knowledge can then be seen as always starting from and ending with problems—problems of ever-increasing profundity and ever-increasing stimulation to tackle new problems.

Requirements for progress of science

Popper suggested three requirements for the growth of scientific knowledge. The first requirement is that a theory should proceed from some simple, new, and powerful idea about a connection between previously unconnected observations. The second requirement is that the theory should be independently testable. In addition to explaining all the previous observations that it was designed to explain, it must be able to predict phenomena that have not yet been observed. This requirement was considered to be indispensable because it is always possible to propose a theory that fits a given set of observations. This second requirement will also serve as a guide for exploring new territory. It means that it will suggest new experiments and, even should these result in a refutation of the theory, they may lead to unexpected results or further experiments that will advance our knowledge. The third requirement is that the theory should be successful in passing some new and severe tests. This is different from the second requirement in that it not only makes predictions about results that can be refuted but that the theory resists the attempted refutations. If the progress of science is to continue, we need not only refutations but also successes. That is, we need to occasionally produce theories that are successful. An unbroken sequence of refuted theories would leave us bewildered and with a feeling that we are not getting closer to the truth.

Dangers to progress

Is there a danger that the growth of scientific knowledge will come to an end because science has completed its task? Popper believed the answer to be an emphatic no due to the infinity of our ignorance. Rather than completing the task, he saw other dangers to the growth, of which he specifically referred to three.

The first is a lack of imagination, sometimes due to a lack of real interest. Scientists develop a curiosity about the world and this serves as the motivation for pitting one's intellect against the unknown in order to discover new things. If this spirit of inquiry is absent, there is no motivation for carrying out a research program. Throughout history, there have been periods where this motivation has dwindled and, as a consequence, scientific progress has waned and even come to a halt. Humans who live under great hardship and have to spend their lives struggling to survive have little scope for intellectual curiosity. At the other extreme, those who live in excessive affluence may not be prepared to step out of their comfort zone to wrestle with unsolved problems. Scientific research is a very exacting profession requiring discipline, dedication, and mental stamina. It often leads to periods of frustration although, if these are overcome, the rewards can be great.

The second danger that Popper foresaw was a misplaced faith in formalization and precision, something which he does not expand on in much detail. As mentioned earlier, progress in science does not mean a continual accumulation of observations but a critical analysis of theories and their refinement or replacement by better ones. Thus, there is a danger of scientists attaching too much importance to observations and their precise measurement while ignoring the theoretical concepts that are needed to explain them.

The third danger to progress in science perceived by Popper is authoritarianism. Throughout human history, there have been many examples of this. Those who questioned that the earth was the center of the universe were persecuted. The science of genetics in the Soviet Union was dominated for several decades by a dogma called Lysenkoism, which was imposed so as not to conflict with political ideology. In such authoritarian regimes, dogma prevails and scientific progress tends to stagnate. A form of authoritarian influence that was not so evident in Popper's day is the managerial systems that have been imposed on science in recent times. These systems with their hierarchical structures and command and control mechanisms will be discussed more in later chapters.

References

Feyerabend, P.K. 1975. *Against Method*. New Left Books, London.
Hanson, N.R. 1958. *Patterns of Discovery: An Inquiry into the Conceptual Foundations of Science*. Cambridge University Press, Cambridge.
Kordig, C.R. 1975. *The Justification of Scientific Change*. Springer Netherlands.
Kuhn, T.S. 1962. *The Structure of Scientific Revolutions*, 3rd edition. The University of Chicago Press, London.
Popper, K.R. 2002. *Conjectures and Refutations*. Routledge, London.
Toulmin, S.E. 1953. *The Philosophy of Science: An Introduction*. Hutchinson's University Library, London.

chapter three

How the lack of scientific input impacts research organizations

If science can provide a more balanced and objective way of thinking than what is used in much of the decision making that is currently made, why is it that very few scientists hold positions of influence in public life? There are, of course, notable exceptions. National leaders such as Margaret Thatcher (Great Britain) and Angela Merkel (Germany) had backgrounds in science. Government leaders in China and Singapore have included high proportions of scientists and engineers in recent times. In contrast, Western countries have had a dearth of scientists within governments. Perhaps the most stellar example of a dual career in science and politics is that of Benjamin Franklin (United States). Franklin, not merely had a background in science but made distinguished contributions to both science and politics. Fernando Henrique Cardosa was a scientist of social sciences and served as president of Brazil for two terms. After retiring from political life, he returned to lecture at Brown University in the United States. Despite the individuals mentioned, the numbers of scientists who venture into public life are few and far between. It has been reported (Paulos, 2012) that out of the 535 members of the U.S. Congress, only 22 had science or engineering backgrounds and, of these, only 2 might be considered as experienced scientists or engineers.

Australian ministers in charge of science portfolio

To emphasize the point that scientists are conspicuous by their absence from public decision making, let us take as an example the credentials of federal ministers for science in Australian parliaments in the past eighty years or so. There have been thirty-one ministers for the portfolio that has included science in the period 1932 to 2013 (*Wikipedia*, 2016). The name of the relevant ministry has changed over this time. In 1932, the title was Minister in charge of Development and Scientific and Industrial Research. In 2013, it had changed to Minister for Tertiary Education, Skills, Science and Research. In 1988–1990, the title was Minister for Science, Customs and Small Business, and in 1996–1998, it was Minister for Industry, Science and Tourism. This lumping of science with other portfolios suggests a downgrading of the perceived importance of science. Of the thirty-one ministers during the years 1932 to 2013, seven were graduates in law, most of whom had practiced in their profession, five were economics graduates, and two had combined degrees in law and economics. None were scientists, although a few had some technical background. This is not to suggest that scientists have been overlooked in assigning the science portfolio. The truth is that very few scientists are available, simply because hardly any enter politics and we will look into why this is so a little later. It does, however, illustrate the point that important decisions about scientific matters are not generally made by scientists but are left in the hands of other professionals. Some of the repercussions of this are discussed later in the chapter. One can only wonder what would be the reaction if, for example, attorneys general were selected from the ranks of scientists.

The contribution of the Honorable R.G. Casey

Of course, to become a successful minister of science does not necessarily require formal qualifications in science, although an understanding of science seems desirable. There have been good ministers of science in Australian governments. R.G. Casey was a graduate in engineering and worked for a time as a mining geologist before entering political life. He was the minister in charge of the Commonwealth Scientific and Industrial Research Organisation (CSIRO) of Australia in the early period after its formation (1950–1960) from its precursor the Council for Scientific and Industrial Research (CSIR). This was a golden age for the organization when many important scientific advances were made that greatly benefited the nation's industries and economy. Casey was held in high regard by CSIRO personnel. He was tireless in striving to find financial support for scientific researchers. He and the government he served in were strong supporters of scholarship and science, and this provided an environment propitious for stimulating creativity. Casey not only had a passion for science but he also developed personal contacts with influential people both in Australia and abroad. This led to acquiring financial support for new expensive projects from organizations such as the Carnegie Corporation and the Rockefeller Foundation. Lord Casey, as he became, nurtured excellent rapport with research scientists of CSIRO. This enabled him to acquire a strong understanding of their research objectives and to be able to effectively communicate their projects to his peers in government so as to gain support for apportioning of resources. It was not accidental that a period of expansion in CSIRO accompanied by outstanding scientific advances coincided with the presence of a minister who understood science and who strove to support the endeavors of the organization. Contrast this with other periods. For example, for another ten-year period from 1988 to 1998, there were seven different federal ministers with responsibility for the science portfolio. This lack of stability again is suggestive of a carefree attitude by successive governments toward science as a vital national activity.

A short history of CSIRO after its formation

In recent decades, there has been a shift in the way scientific research has been organized. In order to illustrate this, we will take as an example what has happened in Australia and will focus on the changes that have occurred in its premier government scientific body, the CSIRO. The CSIRO was established under an Act of Parliament in 1949. Its structure was made up of divisions and smaller groups called units, each dedicated to a specific industry. At that time, wool and minerals were among the most important industries. There was also work being carried out in fundamental areas of research such as in the divisions of National Standards and Radio Physics. The organization was headed by an Executive Management Council that consisted of a full-time chairman and four full-time members, all of whom were outstanding scientists and employees of CSIRO. In addition, there were a small number of part-time members who were drawn from business and university backgrounds. The divisions were headed by chiefs who were eminent scientists in their fields. The chiefs and senior scientists were well supported by administrative staff so that they could devote their efforts to directing the research. Members of the Executive were also in close contact with scientists at the bench and kept well informed by regular visits to laboratories. Research scientists were relatively free to pursue their goals in accordance with the aims of their division and in consultation with their chiefs. Most held positions of tenure and were not under great pressure to seek external funding.

Changes resulting from reviews of CSIRO

The structure of CSIRO was to change as a result of two major reviews, one in 1977 and the other in 1986. In 1977, the committee of the Birch Review, headed by one scientist and two representatives from industry, organized forums at which scientific staff were invited to provide input. Implementation of its final report did not greatly affect the way research was conducted. The main change that was introduced was that a new tier of management was inserted between the divisional chiefs and the Executive Council. This involved a level of Institutes, which consisted of groups of Divisions in related areas of research, each headed by a Director. The role of the Institute Director was to be managerial (CSIRO Annual Report, 1977/78).

In 1986, the global management consulting firm McKinsey and Company was contracted by the government to carry out a review of CSIRO. No forums were arranged to canvas feedback from scientists, and consultations were only held with senior and selected members of the organization. The recommendations following this review were mainly adopted. The result was the most far-reaching restructuring of CSIRO in its history. The Executive Management Council, which had consisted of outstanding scientists, was abolished and replaced by the CSIRO Board, which assumed office in 1987. The first Board consisted of ten members, headed by a chairman, who was a retired politician. The Board included four company directors, a professor of economics, and a trade union official. The scientist who had previously held the position of Chairman became a member of the board and was designated as the chief executive. The Chief Executive was now responsible for the management of the organization in accordance with the strategy, plans, and policies approved by the CSIRO Board. The Board was directly responsible to the Australian government for the overall strategy, governance, and performance of CSIRO. Although some changes had been put in place as a result of the 1977 review, it was the 1986 McKinsey review that was to fundamentally alter the direction and culture of the CSIRO. Before discussing these changes, it may be useful to expand on the nature of scientific research that was touched on in Chapter 2.

How scientific knowledge is acquired

In order to make it easier to assimilate for the nonscientist readership, a very simplified version of the way scientific research proceeds will be given. Let us do this by first considering how some of the great scientific discoveries have been made that have had profound impacts on civilization.

Albert Einstein worked as a technical officer in the Swiss Patent Office. During a short time spent there, he made some of the most important discoveries about the physics of the universe. It is believed that his work at the patent office did not require great mental effort and this allowed him to dedicate his time to pursuing problems that challenged his intellect. One of these was what today we refer to as the theory of relativity. Einstein showed that when problems are tackled that involve large distances and high velocities (such as the velocity of light), it is necessary to include the dimension of time and to replace 3-D space by a 4-D space–time continuum. Some of the consequences of the theory were that light should be subjected to gravitation and that a moving clock runs more slowly in a gravitational field and is shortened in length in the direction of motion, a result that appeared counterintuitive to thinking at that time. Another result was that mass and energy were interconvertible and their relationship was governed by a simple equation, $E = mc^2$, where E is energy, m is mass, and c is the velocity of light, the equation that explains nuclear reactions.

Charles Darwin made use of his journey on the *Beagle* to make observations and collect specimens. Study of the specimens together with avid reading challenged his mind to try to explain how different species evolved, leading to his theories of evolution.

Irving Langmuir was an American industrial scientist who worked for the General Electric Company. A problem that was faced at the time was the short life of the recently invented tungsten filament lamp, caused by the rapid blackening of the bulb. Langmuir made an intensive study of the fundamental science involved, including the effects of surrounding the filament with different gases. This led to the use of the inert gas, argon, which overcame the problem and increased the lifetime of the filaments. The knowledge that was acquired from these studies led to applications in other areas such as the development of high vacuum tubes, the isolation of atomic hydrogen, and the process of arc welding.

Louis Pasteur studied fermentation at the University of Lille. A problem at the time was the spoilage of beverages such as beer, wine, and milk. Pasteur showed that when particles were prevented from entering flasks containing broths, no spoilage occurred, showing that spoilage must occur through organisms entering from outside. This work led to the use of antiseptics and the development of vaccines for different diseases.

How managerial control purports to drive science

In many scientific organizations these days, the performance criteria for technical and research staff are formulated by administrators. At the start of the review period, they are required to document such things as objectives, tasks to be performed, milestones to be accomplished, time lines to achieve objectives, expected outcomes, and financial benefits expected to accrue. At the end of the review period, the milestones and expected outcomes are assessed against the initial objectives. Some flexibility may be incorporated, but this rather rigid framework conflicts with the way that scientific research should proceed.

Another way of illustrating the managerial approach to research is to examine how proposals for financial support from funding bodies are required to be prepared. These have been very well summarized by the Nobel laureate Albert Szent-Gyorgyi (1974), who wrote, in relation to grant proposals to the U.S. National Institutes of Health (NIH):

> Research means going into the unknown, which demands a pioneering spirit. This spirit is now strangled by the way in which the main biomedical granting agency (NIH) distributes its grants. The unknown is unknown because one does not know what is there. If one knows what one will do and find in it, then it is not research any more and is not worth doing. The NIH wants detailed projects, wants

applicants to tell exactly what they will do and find during the tenure of their grants, which excludes unexpected discoveries on which progress depends.

Comparison of the two approaches

The intention has been to contrast the two different strategies for carrying out scientific research. These will be denoted as the scientific approach and the managerial approach. The scientific approach is illustrated by the four examples described earlier. Let us examine these examples to try to throw light on the essential features and how it differs from the managerial approach. Albert Einstein did not set out to discover the theory of relativity. This theory was unknown when he began to challenge his mind with the questions that were presented to him, based on the observations and speculations of previous researchers. It is inconceivable that he could have formulated objectives and time lines to achieve them. It is only possible to do this when the objectives are mundane and relatively trivial. Important scientific discoveries are made by individuals who acquire advanced knowledge and are fully conversant with what has been found by previous workers in the area of investigation. Then, armed with this knowledge, they relentlessly pursue ideas wherever they may lead, until insights into previously unimagined territory may be revealed.

This is the pattern that is evident in the research of the four scientists considered earlier. When Charles Darwin began his studies of the different species that he encountered, he did not have as an objective to develop a theory of evolution. He set no milestones or time lines to achieve them, although, in hindsight, it is possible to recognize that there were milestones. The initial efforts of Louis Pasteur were not to develop vaccines that would revolutionize the treatment of diseases because the concept of vaccines was not properly understood. Irving Langmuir did not directly set out to invent viable electric lamps, although this problem was always at the back of his mind. By studying the fundamental science involved, he was able to open windows of opportunities that had not previously been conceived.

In contrast, the setting of objectives and times to achieve them, as is done in the managerial approach, limits discoveries to only what was originally planned. This type of research is therefore destined to be pedestrian. It eliminates the windows of insight that open up when scientists have the freedom to pursue their natural spirit of inquiry, unshackled from constraints imposed by management. In this latter activity, scientists frequently have to go along dead-end roads. This is not understood by management who are obsessed with efficiency. However, as we have seen from what was described in Chapter 2, this is an essential component of research. In order to progress, scientists need to make mistakes and to learn from them.

How has the CSIRO fared?

Returning to our discussion of the CSIRO, implementation of the recommendations introduced following the McKinsey review in 1986 has produced a change from a bottom-up organization to a top-down one. What this means is that instead of research scientists having a good deal of freedom to go about their inquiries, as had existed before, they were now constrained more by the need to comply with management directives. In other words, the culture began to change from a scientific approach to a managerial approach. The structural hierarchy began to undergo changes. In 1990, the term of the Chief Executive, who was an eminent scientist, concluded. Since that time, the selection of the Chief Executive has required the appointees to have science backgrounds but also to have management experience. This has meant that the appointees have not needed to be world-renowned research scientists but needed to have managed scientific groups. The proportion of managerial positions relative to research positions in the organization has steadily increased and many of the managers are on higher salaries than the active research scientists. Whereas, in former times, scientists had a good deal of freedom to develop their research, a system has developed with different levels of management. Divisions are organized into programs that, in turn, are divided into subprograms and then into projects, each level having a designated leader (perhaps more appropriately referred to as a manager).

The requirement to obtain a proportion of the research funding

Another outcome of the McKinsey recommendations was that divisions should take responsibility for obtaining a proportion of their funding from external sources. This requirement has a number of consequences. It means that scientists have to spend an appreciable amount of their precious time preparing grant applications and writing reports for the funding bodies. The essential feature of these grant applications is that they are necessarily for short-term projects, usually for a maximum of three years. In former times, scientists worked on long-term projects. This enabled them to build reputations as world-leading researchers in the particular field and placed them in a strong position to be able to carry out research of high quality for the public good. The transition to short-term projects meant the need to formulate objectives that could be achieved in a short time. Thus, the quality of the research was diminished. CSIRO changed from an organization devoted to long-term research for the public good to what was essentially a corporatized body dealing with short-term consultancies.

Effect of these changes on the workplace environment

The imposition of managerial policies on research scientists has resulted in a loss of morale and job satisfaction and a steep increase in stress accompanied by health problems (Rees, 1995). Scientific research is a creative activity in which researchers need to have a degree of tranquility to allow their natural spirit of inquiry to pursue the challenges that they confront. The managerial ideology does not recognize this. The basic tenet of managerialism is that the performance of all organizations can be optimized by the application of generic management skills and theory. According to this premise, there is little difference in the skills needed to run a research establishment, an advertising agency, or a factory turning out appliances. The experience and expertise associated with the core business of the organization are of secondary importance. To be treated on the same basis as a factory employee working on a production line is demoralizing for a research scientist.

The effect of supplanting long-term research by short-term projects is to threaten the employment stability of research scientists. To obtain a quick-fix to some relatively trivial problem that might only take a short period may not require researchers with a base of knowledge and experience that has been built up over a long time. This means that there can be a shift from indefinite appointments to term appointments, giving the organization a greater flexibility to terminate research staff and make appointments for fixed terms. This is a recipe not only for a lowering of morale but a decline in the quality of research. Another factor that has a negative effect on the quality of research is the requirement for scientists to procure an appreciable portion of external funding. This can be of the order of 30 percent of the total cost of their projects or more. We have already considered the inefficient waste of all scientists' time in preparing research proposals and reporting to funding bodies. But there are other consequences that are detrimental. In the case of "public good" research, the funding has to come mainly from other federal and state departments (Paltridge, 2012). As a result, proposals for funding are targeted toward issues that are currently fashionable to politicians. Whenever there is a change in direction of government priorities, additional new research staff are appointed, leading inevitably to an increase in the salary bill and making the employment of all staff more insecure. Predictions of massive job cuts in CSIRO in recent times have been reported in the media (*The Sydney Morning Herald*, April 6, 2016).

The climate change controversy

One of the current fashionable issues is that of climate change. There is a widely propagated view that there is an appreciable contribution to warming of the planet by an increase in greenhouse gases in the atmosphere, resulting from burning of fossil fuels. This causes the formation of a kind of blanket that reduces the escape of radiant heat, leading to warming. The science behind this has been mainly based on theoretical modeling rather than experimental testing. An international body, the Intergovernmental Panel on Climate Change (IPCC), has been formed from scientists and bureaucrats, and its regular reports have had an enormous influence on public opinion. The IPCC has been structured into three working groups. The first deals with the science behind the predictions of global warming. The second deals with the potential impact of warming on human society. The mission of the third has been to examine and recommend options for international action to avoid the assumed disastrous consequences of climate change. The third group has been the most influential. It has not waited for input from the other two groups. Rather, it has moved on the assumption that disastrous global warming is real and its effects need to be urgently addressed.

The lack of scientific thinking in the debate

It is not the intention in this book to enter into opinion for or against whether there is an appreciable anthropogenic effect on global warming. The discussion will be confined to evaluating the scientific thinking that has been applied to the issue and to what have been the consequences. The first thing to note is that there is complete dissension between those who believe that there is significant anthropogenic global warming and those who do not. Of course, dissension is an important component of scientific debate. When scientists argue about issues, it is usually positive, provided they are prepared to put their point of view but to also be disposed to consider opposing views. This is not the case with the climate controversy. Those who support the theory of warming are called "alarmists," while those who oppose it are designated as "deniers." The idea is that by denigrating their opponents (i.e., those with the contrary view) it is inferred that their opinions are not worth considering. This is not how science is supposed to work.

Errors in application of science

Of course, many of those who participate in the debate are not scientists, and many of those who have scientific credentials and call themselves scientists do not adhere to scientific principles. The debate has become more political than scientific. There are those who are activists for various causes and use the debate to push their agendas. As a result, truth gets moved aside. Another mistake that has been made is to assume that the opinion of the majority of scientists should be accepted. In science, consensus counts for nothing. In fact, most of the great discoveries of science have been made by those who have not followed consensus. One often hears statements such as "the science is in" or "it has been scientifically proven." Such statements are erroneous and show an ignorance of science and its philosophy, the topic that was discussed in Chapter 2. Another common mistake is to attach unjustifiable significance to observations that appear to be consistent with a particular theory. It is easy to obtain confirmations if one looks for confirmations. This is the dogmatic approach, a tendency to try to verify our beliefs even to the point of neglecting observations that conflict with those beliefs. The true scientific method, as discussed in Chapter 2, introduces a critical attitude that is prepared to subject a theory to possible falsification or refutation.

Balanced debate can still occur

Despite the mire that the controversy has spawned, there have been contributions to the debate that have been refreshingly balanced and scientifically acceptable. One example is the monograph by Paltridge (2010). Here, the author has no agenda other than to examine the issue in an objective way. He acknowledges that the basic premise of the global warming theory is correct, that is, that concentration of carbon dioxide in the atmosphere should lead to warming. However, he argues that, because of the complexity of the problem and its associated uncertainties, it is not possible to determine how large the effect could be. He well summarizes the history of the controversy and explains how different influences, especially that of the IPCC, have created a juggernaut. A politically correct point of view has emerged that anthropogenic global warming has been proven and those who question it, of whom there are many, are finding it difficult for their voices to be heard. Those who submit papers to scientific journals that question the popular view find themselves on a playing field that is not level. They are subjected to greater obstacles than those whose papers are in accord with the prevailing theory.

Expansion of the managerial approach

In recent years, the culture in CSIRO has moved toward more managerialism. There has been a massive reorganization of the research structure and external communication. "Flagships" have been introduced. These are programs that are designed to focus on Australia's most significant challenges and opportunities and are apportioned additional funding. They are superimposed on the traditional discipline-based divisional structure, the profiles of which have been greatly reduced. Flagship clusters have been formed in which the CSIRO forms partnerships with other research bodies (Australian and overseas institutions, Cooperative Research Centres, and at least one Australian university), which are required to provide matching funding. Research is managed by what is called a matrix system in which there are various streams through which scientists report to different managers. This system has been strongly criticized for placing emphasis on management rather than on research and the scientists who carry it out (Paltridge, 2012; Pockley, 2013).

The present working environment

One of the flagships that was mentioned in the previous section is climate adaption. The mission of this flagship is to have a particular focus on better understanding and preparing for the impacts of climate change. Top management in CSIRO adopted the view that anthropogenic global warming has been proven and its effects, unless addressed, will be disastrous. The result is that there is pressure on scientists to channel their research to accord with what has become the prevailing view. Conversely, it is becoming more difficult to carry out research that could challenge what has become the politically correct position. This is not the way that scientific research is meant to proceed. In relation to this, writer and political commentator Don Aitkin (2012), who was appointed as the first chairman of the Australian Research Grants Committee, wrote the following, referring to what scientists call postnormal science:

> To me, this has been the worst outcome of the move into "priorities": research has become an adjunct of politics. You can see this most clearly in "climate change." I read through the long list of project titles in that field supported by the Australian Research Council, and it seemed to me that most if not all seemed to take anthropogenic global warming (AGW) for granted, even though we still cannot distinguish the "signal" of AGW from natural variability.

The Victims of CSIRO blog site

As mentioned earlier in the chapter, signs of demoralization in CSIRO with accompanying effects of stress and health problems were documented in the 1990s (Rees, 1995). In recent times, the work environment for scientists has progressively worsened. It is not a coincidence that these signs began to intensify with the advent of the introduction of new management controls on research. For those who understand science, these effects are not unexpected. 1986 was the year that the McKinsey recommendations were adopted. A blog site titled "Victims of CSIRO" (https://victimsofcsiro.com) was formed in mid-2011 by current and former employees who have been subjected to bullying and other misconduct. The concerns of these scientists have been summarized in a series of blogs on the Victims of CSIRO site. More than 60 scientists have claimed to have suffered bullying and intimidation.

References

Aitkin, D. 2012. Science and politics: Who pays the piper? http://donaitkin.com/science-and-politics-who-pays-the-piper/. October 31.

Paltridge, G.W. 2010. *The Climate Caper*. Taylor Trade Publishing, Lanham, Maryland.

Paltridge, G.W. 2012. Has the CSIRO lost its way? *The Australian Financial Review*, October 19.

Paulos, J.A. 2012. Why don't Americans elect scientists? *New York Times*, February 13, The Opinion Pages.

Pockley, P. 2013. Science lost in CSIRO's matrix. *Australasian Science Magazine*, May. www.australasianscience.com.au/article/issue-may-2013/science-lost-to-csiros-matrix.html.

Rees, S. 1995. The fraud and the fiction. In *The Human Costs of Managerialism: Advocating the Recovery of Humanity*, S. Rees and G. Rodley, eds., 15–27. Pluto Press, Sydney, Australia.

Szent-Gyorgyi, A. 1974. Research grants. *Perspectives in Biology and Medicine* 18:41–43.

Wikipedia. 2016. Minister for Industry (Australia). Accessed April. http://en.wikipedia.org/wiki/Minister_for_Industry_(Australia).

chapter four

How could this have happened?

How could it have happened that a scientific organization (Australia's Commonwealth Scientific and Industrial Research Organisation [CSIRO]), renowned throughout the world for its research excellence, has transformed into one with an allegedly toxic workplace environment and a blemished reputation? This is the question that we will attempt to come up with answers to in this chapter. To throw light on the question, we will first examine how a career in science differs from some other careers.

A career in science

First, a career in science is a career in itself. Scientists are motivated by a spirit of inquiry. To become a successful research scientist requires years of study and application. Discovery of substantial new knowledge is an arduous task. Frequently, recognition of success comes only toward the middle or end of scientists' careers when they have had a significant impact. Thus, they have a lot invested in their careers and are reluctant to leave before they have made an appreciable contribution. In contrast, a career in law or economics is often embarked on with the view that it will be a stepping-stone for entering public life. Other professions that frequently feed into politics are journalism, public relations, school teaching, and business administration. After serving for a relatively short time in their profession, these specialists may be quite happy to step aside and seek a political career. Even if they are elected and later voted out of office, it will not be so difficult to resume their former career or another related one. The same is not the case for a scientist. A successful career in science usually requires a lifetime's dedication to build the expertise needed to be able to push forward the boundaries of knowledge. Periods of absence cause research scientists to fall behind in their awareness of how their field has progressed and this cannot easily be recovered. To carry out worthwhile research requires the stability of employment that allows scientists to focus their undivided attention on the problems they address. This is one of the factors that has caused lowering of morale of scientists when the security of their jobs is threatened. It is what has happened to the scientists of the CSIRO as was discussed in Chapter 3. In the decades from its formation in 1949 up until the mid-1980s, research scientists enjoyed a high degree of security in their employment. After joining the organization, many could count on continuing their research careers until retirement if they wished. This enabled them to make significant contributions for which many attained international recognition. There has been a change of the status quo from the late 1980s to the present. Research for long-term public benefit has, to an increasing extent, been supplanted by short-term projects and consulting. Topics that receive support for longer-term research are increasingly restricted to those that are politically correct such as that of climate change.

Effect of the requirement to procure funding

Another source of demoralization and one that is connected with the uncertainty of permanent employment has been the introduction of the requirement for scientists to procure a large proportion of their funding from external sources. This has meant that either they seek funding for short-term projects to be supported by industrial companies or they need to attach themselves to research that is of a "public good" nature but may necessitate having to conform to political correctness. Both of these alternative avenues lead to a dumbing down of the quality of research with a corresponding lowering of morale. The short-term projects do not necessarily need researchers who are established world-class scientists. It therefore opens the way for new appointees to be brought in. The effect is to threaten the employment security of all scientific staff. In the case of long-term projects, many may not be prepared to sacrifice their scientific integrity to work on what may be perceived as being driven by politics rather than science. As a result, "Top flight researchers have departed to find scientific freedom elsewhere while others have been pushed aside" (Besser and Phillips, 2013). For those who do not adequately understand the nature of scientific research, it probably seems sensible that scientists be made responsible for a large proportion of their funding. The negative consequences of this requirement nevertheless become evident with time.

Why do not more scientists enter politics?

This is a question that has attracted some attention and is a good question to illustrate how scientists differ from other professionals. The skills required to become a successful politician are different than those required to be a scientist. In a democratic electoral system, the success of a politician does not so much depend on their ability to formulate policies for the long-term benefit of society. It is more about manipulating public opinion to accord with the policies of the party to which they belong. They also need to attack the policies of their opponents. Lawyers fit well into the political system, as they are often experienced in adversarial skills. Scientists can also be adversarial when they debate the evidence for or against a particular theory. It is, however, different to politics. A scientist (or at least one who can honestly claim to be a scientist) will weigh the evidence in a debate and will readily admit if they feel that the opposing argument is superior. After all, the object of science is to discover the truth. If a politician were seen to frequently admit they were wrong, it is unlikely that their career would prosper.

Are scientists different?

There is a widely held view that scientists are different in personality from nonscientists. This belief has strong support from an extensive study of a large group of gifted people over a period of more than thirty years (Terman, 1955). The study was undertaken with two main objectives. The first was to identify potential scientists at an early age and the second was to try to identify some of the human factors that contribute to misunderstandings between scientists and nonscientists. The study began in 1921 and the original group consisted of 800 males and 600 females who were selected from the top one percent in intelligence of the school population, measured by mental tests. For the comparative study of scientists and nonscientists, the study was confined to 800 men, as only a few of the women pursued scientific careers. The careers of the 800 men were followed in a continuous manner by questionnaires mailed to them and by four field studies in 1921 (when they were children), 1927, 1939, and 1950.

The group was divided into seven subgroups. The first subgroup (physical science research, total 51) comprised research workers in basic physical science or engineering. The second subgroup (engineers, total 104) was practicing engineers who had done some applied research. The third subgroup (medical-biological research workers, totaling 61) were workers in biological research or medicine. The fourth subgroup (physical or biological science, nonresearch, totaling 68) comprised men who had majored in a science but had mainly gone into other fields of work, such as teaching. The fifth subgroup (social science, total 149) had majors in social science and most were in business occupations. The sixth subgroup (law, totaling 83) were lawyers. The seventh subgroup (humanities, total 95) were men who had majored in a field of humanities; this subgroup had gone into a variety of occupations, mainly teaching and business.

Differences in traits among the groups were investigated under several hundred variables. The chi-square (square of the correlation coefficient) was used to measure significant differences between the subgroups for different variables. Variables were grouped into three categories: (1) scientific interests evidenced, (2) interest in business occupations, and (3) social traits and social adjustments. For the first category, the physical scientists, engineers, and science majors scored high to very high on most variables. The nonscientists scored consistently low. For the second category (interest in business occupations), the nonscientist group scored highest, the three groups of research workers in science scored low, and the science majors (nonresearch) were intermediate. In the case of social traits (third category), lawyers and social science majors rated highest; physical science researchers, engineers, and science majors rated lowest; with the medical-biological and humanities majors intermediate. The clear result of this detailed study of a gifted group of people showed that physical scientists and engineers are at the opposite pole from the businessmen and lawyers in abilities, in occupational interests, and in social behavior.

How science of excellence can still be achieved

It should be acknowledged that, despite the obstacles that have been discussed, research of excellence is still being carried out in CSIRO. This may relate more to the indomitable human spirit that has triumphed over adversity throughout history rather than to the controls imposed by management. It is opportune at this time to reflect on one of the important advances that has been made by CSIRO in recent times. This is the invention of wireless local area networking (WLAN), which has become the foundation for wireless networking technology (Wi-Fi). Wi-Fi allows exchange of words, data, and videos with any computer across the Internet or around the world. The technology is in several billion devices such as desktop and laptop computers, workstations, television sets, cameras, and smart phones, and is now used in offices, public buildings, and homes everywhere. It was developed by a CSIRO team led by John O'Sullivan and included Terry Percival, Diet Ostry, Graham Daniels, and John Deane (Moses, 2010).

It is fascinating to recall how this technology developed. It can be traced to 1974 when the theorist Stephen Hawking suggested that, under certain circumstances, small black holes might disintegrate and this would cause them to emit radio signals. These small black holes are believed to be cosmic objects that contain a huge mass that is concentrated in a tiny volume, possibly smaller than an atom. The CSIRO team set out to try to measure the predicted signal. A difficult problem they were met with was that radio waves from an outgoing signal bounce around in an indoor environment causing distortion of the signal. The team members drew on their experience in radio astronomy to solve this problem. It involved using complex mathematics known as fast Fourier transforms together with knowledge of radio waves and their behavior in different environments. Although the original objective was to obtain information about black holes, the relentless search for a solution to an accompanying problem has led to a new technology with a vast application that had previously not been imagined. It was made possible by scientists who possessed advanced fundamental knowledge of their field and the spirit of inquiry and persistence needed to resolve the obstacles that they encountered.

How could it have happened?

We return to the title of the chapter and the question that was asked in the first sentence. Some have tried to answer the question but I believe we need to go to 1986 to find the major turning point. It was then that the Australian government decided to appoint the global management firm McKinsey and Company to carry out a review of CSIRO to look into its management structure. The background to this decision has not been clearly documented. There have been suggestions that the government was not happy with the amount of fundamental research and was finding it difficult to trace the expenditure and relate it to the outcomes of the research. It may also be relevant that a member of the CSIRO Board was a former consultant and director of McKinsey.

Making decisions without the basic knowledge

In retrospect, it seems remarkable that a decision was made by the government to appoint a management company to undertake a review of a scientific research organization. Obviously, there was a lack of understanding of how science works among government ranks. As has been noted, there are few in government with a technical background let alone any with scientific research credentials. Professions such as law, economics, and business management are the ones that predominate. This is reflected in the fact that the company chosen to carry out the review had only management expertise and there was no input from science. The basic premise of managerialism, as already noted, is that all organizations have more similarities than differences. According to this, the performance of organizations can be optimized by the application of generic management skills and theory. For those practicing this system, there is little difference in the skills required to run a research body, an advertising agency, or a factory turning out machines. The experience and skills associated with each organization's core business are considered to be unimportant or, at least, of secondary importance. This is consistent with one of the recommendations that was made and accepted. This was that a vigorous program of management training should be introduced (CSIRO Annual Report, 1986/87, p. 7). We need to go little further than this recommendation to understand how the culture of the Organization was to change and how the quality of its research and the morale of its scientists were to undergo a downward spiral. The conclusion that we can take from this is that management structure was considered to be of greater importance than the science.

How important scientific discoveries are made

The way that scientific discoveries are made was outside the comprehension of the Australian government as well as the management company that it contracted. The development of Wi-Fi, discussed earlier, illustrates how great scientific discoveries are made. The team that developed the technology worked for years on the problem without an outcome. Its advanced fundamental knowledge combined with relentless pursuit of ideas eventually triumphed. It was fortunate that the team worked in a CSIRO Division (Radiophysics in those days) that supported and encouraged fundamental research. If it had been under the control of a managerial hierarchy that saw little merit in fundamental research, it is likely that the team's work in the early years would have been considered unproductive and the project may have been terminated prematurely.

Effect of organizational changes on research

It will be interesting to see how historians record the evolution of the CSIRO from an organization headed by outstanding scientists backed up by administrators to what exists today. After 1986, the number of divisions was reduced by lumping together smaller divisions and units, often from different locations in the country. The role of the Divisional Chiefs, who were eminent scientists, was thus diluted. They now had to devote more of their time to management rather than having a direct influence on the science. In earlier days, research scientists enjoyed a certain autonomy. They were able to pursue their investigations, guided by their Divisional Chiefs and stimulated by colleagues who were working in similar areas. This has changed to a system where research scientists have lost a good deal of their independence and are subjected to the control of lines of management.

Scientists now have to compete with each other to procure a substantial amount of their funding. This has eroded the collegiality that existed previously as they now have to "learn to claw at each other" for the scarce funding (Besser and Phillips, 2013). The matrix management system, which was mentioned in Chapter 3, is one in which scientists report to a series of managers through different streams. The organization now appears to have management as its core activity rather than science. As a result, some scientists, in order to guarantee their tenure, try to move into management positions, as that seems to be the path to promotion. It is small wonder then that a toxic work environment has formed in which bullying is claimed to be widespread.

Where angels fear to tread

The answer to the question posed by the chapter title may be tentatively summarized as follows. The decision to contract a business management company was made by politicians who had little understanding of how science operates. They were under the illusion that scientific research could be made more relevant to the country's needs (i.e., more of an applied nature) and the scientists more accountable by imposing stricter management control. This failed to take into account that most important scientific discoveries emanate from fundamental research, and it is vital to maintain a good proportion of this type of research. Scientific research is a creative endeavor. It cannot be supervised by the same controls that are used to optimize the efficiency of a factory production line and it cannot be "managed." The decisions made by politicians that have led to the erosion of a prestigious research group have displayed a failure to understand the scientific method and an absence of critical thinking.

Where was the scientific leadership?

We should not be too harsh in our assessments. The decisions made by the politicians were made with the best intentions. Similarly, the recommendations made by the management consulting firm were made based on its best knowledge of how organizations should be structured. Unfortunately, it was assumed that a scientific organization does not have any unique characteristics. The contribution to the decision making from science was minimal. As previously noted, scientists are largely absent from politics so that there were no scientists in government who could be called on for advice. Scientists tend to be a bit unworldly. They are not usually noted for their assertive nature, at least in public. It is nevertheless somewhat puzzling that there was little outrage expressed by CSIRO staff, particularly those scientists in the most senior positions, to the proposed changes that resulted from the review. This raises questions about how well these senior staff members understood how science works and how passionate their willingness to defend its principles. With the wisdom of hindsight, it is evident that these were bone-headed decisions. There was a stark contrast with the decisions made by the government in 1949 when the CSIRO was established. That government recognized the logic of having outstanding research scientists at the top and the importance of allowing individual researchers a good deal of autonomy in pursuing objectives that were vital for the nation.

Why do toxic work environments form in science organizations?

The negative effects of managerialism on organizations are not restricted to scientific ones. However, there are special features of scientific organizations that make them particularly vulnerable. Hierarchical structure is alien to a scientific culture. Albert Szent-Györgyi, a famous scientist who was a Nobel laureate, stated that "the real scientist ... is ready to bear privation and, if need be, starvation, rather than let anyone dictate to him, which direction his work must take."[1] This statement epitomizes the nature of scientific research. It requires a certain degree of freedom to operate successfully. It is a creative activity. Can anyone imagine what would have been the outcome if the great composers and the great painters had been subjected to line management and told how they were supposed to operate? Of course, it has to be recognized that any system can be abused. It may happen that some scientists can enclose themselves in ivory towers and become absorbed in work that is unlikely to lead to useful outcomes. This, I believe, is not very common. In the main, scientists are responsible and dedicate themselves to work on solutions to the problems to which they are assigned. Where difficulties can arise is when they are required to communicate their work to those who make the decisions and who do not have a good understanding of the scientific process.

How do toxic workplace environments arise?

The problems that have led to the formation of the Victims of CSIRO website (mentioned at the end of Chapter 3) are not restricted to scientific organizations. They have also been reported in other areas such as the public service, academia, and hospitals to name a few. A toxic workplace where bullying is rampant causes considerable suffering for many people, not to mention the negative effects it has on productivity. Before attempting to suggest ways to remedy the problem, it is important to understand how it may arise. A top-down line management system is particularly susceptible because it allows certain individuals to have relatively exclusive access to the ears of superiors. There are certain personality types that can utilize such a system for their benefit and to the detriment of their colleagues. We are going to focus on some of these personality types. If they can be identified by a wide cross-section of the workforce and their behavior understood, then this will increase the chance that their effects can be nullified or, at least, minimized. It is only when they are allowed to act clandestinely that they are able to get away with their destructive actions.

The control freak

Control freakery is a personality disorder that appears to be fairly common in the workplace but is particularly difficult to deal with in a scientific environment. This is because independence of thought is an essential ingredient of research. If research scientists have the misfortune to work under a person who subjects them to strict controlling tactics, their independent thinking and thus their creativity is stifled. I will give an example from my own experience. I was successful in being awarded external funding, in open competition, for a project that I had proposed. My immediate manager decided that the project for which I had procured funding should not proceed and put pressure on me to try to get me to divert the funding to be used for a different project of his choosing. Although I resisted, the funding was eventually taken from me and given to another. The control freak, who was my immediate manager, had ingratiated himself with the Divisional Chief and apparently had got his support for the switch. As it turned out, the new line of research, which was funded for three years, led to nothing of any value. This example illustrates how control freaks operate. They are usually people who present a very different picture to their subordinates than what they appear to their superiors. This enables them to move up the career ladder. Those who show independence of thought and do not kowtow to their perceived superiority are seen as a threat to their control and are especially targeted. To their superiors, they show themselves to be reasonable and acting in the best interests of the organization.

Control freaks are usually the cause of micromanagement. They may enjoy the feeling of power they get so much that they try to gain control of everything and everyone around them. They do not see themselves as controlling, simply that they are right.

The narcissist

Narcissism is another personality disorder that causes discord in the workplace. Narcissists have an exaggerated sense of self-importance. They believe that they are superior to others and are unable to recognize other people's feelings. They expect others to go along with their ideas and plans. This is different to people who have a healthy confidence and self-esteem because such people do not value themselves more than they value others. Narcissists have a sense of entitlement and when they are not apportioned the special treatment that they feel they deserve, they may become impatient and angry. Narcissists come in all shapes, sizes, and degrees (Lopez de Victoria, 2008). However, they share certain common characteristics and their effects in the workplace are similar. In a scientific organization in which a managerial culture operates, these people can make it extremely unpleasant for anyone below them in the hierarchy. They persistently, albeit often subtly, exaggerate their achievements and detest those who do not recognize their "superiority." Another characteristic is that they expect automatic compliance since they consider themselves to be special and anyone who fails to comply is branded as deficient.

Narcissists often gain leadership positions because people who are narcissistic have a desire to be leaders. They do not necessarily make better leaders, but because they want to do it, they are more likely to end up in those positions. No one likes being criticized, but people with narcissism are hypersensitive to criticism. They are good at making excuses and not taking responsibility for the mistakes they make. However, they are particularly adept at finding fault. They are good at attaching blame to others and go to great lengths to emphasize the blame.

The serial bully

The Victims of CSIRO website specifically points to bullying, harassment, victimization, and other forms of antisocial behavior in the workplace as the main issues that have prompted formation of the group. Serial bullies are persons who can inflict considerable damage to an organization and leave a trail of human misery. The targets of their bullying are often decent and competent people. The bully may begin by confronting victims on a one-to-one basis to accuse them of some fault. The strategy of the bully is then often to use subtle tactics to put the target in a bad light. This may take the form of lying to their superiors about the target. In a line management system, the bully will have groomed the superiors to believe him/her to be a good person who has the interests of the organization as their primary concern. Once this has been achieved, the superiors are ready to believe all the falsehoods told to them about the target. The bully may also spread subtle lies about the target to the target's colleagues, an activity called mobbing. The target may then find him-/herself in a position where they feel powerless. If the target leaves, the bully then rapidly shifts their attention to another victim. The serial bully will always try to satisfy the desire to show dominance over someone.

The psychopath

All the personalities summarized this far—the control freak, the narcissist, and the serial bully—may be found in a more general personality trait, that of the psychopath or sociopath. The people who are most qualified to discuss psychopaths are those professionals who have made a study of them, which usually means psychologists or psychiatrists. Some valuable authoritative articles on the subject are provided in the references at the end of the chapter. Others who may be able to give insight are those who have been victims of psychopaths and have read extensively on their behavior. Very few victims will be capable of providing useful information because many will have been destroyed by the psychopath and will have finished up severely mentally affected or suicidal. However, there will always be a few with the necessary strength of character to survive and their insights can be valuable.

What is a psychopath?

Very few people understand what a psychopath is. Some confuse the term with psychotic, which is something quite different. Psychosis is a symptom of mental illness and, broadly speaking, refers to someone who has lost contact with reality. Others may associate psychopathy with serial murderers or other violent criminals. It is true that some psychopaths do fit this mold. However, the vast majority give the appearance of being ordinary citizens who may lead apparently successful lives and work in different professions. The jobs that most attract psychopaths are CEOs of companies, lawyers, media personnel, and salespersons (Sheffield, 2015). They may well go through life without anyone suspecting them of having any special personality trait. The important thing for everyone to know is that they are a different type of human being to what most understand human beings to be. Before discussing their characteristic traits, let us examine their frequency in the general population. It is estimated that between 1 and 3 percent of the population are psychopaths but psychopathy is believed to be more frequent in the management ranks of organizations. In fact, Babiak and Hare (2006) suggest that there are some three and a half times more psychopaths in senior positions than there are in the general population.

Characteristics of psychopaths

Perhaps the most distinguishing feature of psychopaths is that they do not have a conscience. Normal people are restrained by their conscience from carrying out acts that might harm others. Psychopaths are not burdened by this constraint. They can use the attribute to great advantage in their daily lives and careers. They can lie and manipulate events so as to cause great damage to others without feeling any guilt or remorse. That is why they are a different type of human being and why such a being is beyond the comprehension of almost everyone. The normal person is so accustomed to having their actions controlled by their conscience that they are unable to conceive of a person who has no conscience. When they see the actions of a psychopath, they think this is just someone like themselves who may be a bit nasty to someone. Psychopaths do not show their true face to the majority around them so they are able to remain hidden. This is particularly frustrating for their victims who often become isolated from the pack in the same manner as a predator animal separates its prey.

How are psychopaths identified?

One of the psychologists to have made an extensive study of psychopathy is Robert Hare. Based on his research with incarcerated offenders and forensic inmates in Vancouver, and helped by the earlier research of Hervey M. Cleckley (2015), he has developed a checklist for evaluating psychopathy (Hare, 2003). This consists of a list of twenty categories of attributes that is scored on a three-point scale according to specific criteria based on file information and an interview. A value of 0 is assigned if the attribute does not apply, 1 if it applies somewhat, and 2 if it fully applies. The maximum possible score is thus 40 and the borderline for clinical psychopathy is estimated to hover around 30. The average score for all incarcerated male offenders in North America was 23.3 in one year. Hare guesses that his own score is about 4 or 5. It should be stressed that amateurs should not use this checklist. Because an individual's score can have serious consequences for people's future, it is imperative that a qualified clinician under standardized conditions administers the test. Otherwise, great harm can result if the test is inadequately administered.

The Hare checklist

Here, we will just pick out a few of the twenty categories from the Hare checklist that will help to illustrate some of the characteristics to be found in psychopaths.

- Look for glib and superficial charm. A psychopath will also put on what professionals refer to as a "mask" of sanity that is likeable and pleasant.
- Look for grandiose self-perception. Psychopaths will often believe that they are smarter or more powerful than they really are.
- Determine if there is pathological lying. A psychopath will tell all sorts of lies, little white lies as well as huge stories intended to mislead.
- Evaluate the level of manipulation. All psychopaths are identified as cunning and able to get people to do things they might not normally do. They can use guilt, force, and other methods to manipulate.

- Look for any feelings of guilt. An absence of guilt or remorse is a sign of psychopathy.
- Consider the effect of emotional response a person has. Psychopaths demonstrate shallow emotional reactions to death, injuries, trauma, or other events that would otherwise cause a deeper response.
- Look for a lack of empathy. Psychopaths are callous and have no way of relating to nonpsychopaths.
- Consider whether the person can accept responsibility. A psychopath will never admit being wrong or owning up to mistakes and errors of judgment.
- Pay extreme attention to the person's treatment toward others. Psychopaths are generally prone to belittle, humiliate, mistreat, mock, and even attack physically (or kill in extreme cases) people who would normally bring no benefits to them in any way, such as subordinates, physically frail or lower-ranking people, children, elderly people, and even animals.

An example of psychopathic behavior

A colleague had the misfortune to fall from a ladder and his injuries forced him to use crutches to move around. Soon after, his boss summoned him to a meeting in which he proposed to accuse him of not deserving the promotion he had recently been given. The colleague had to travel to attend the meeting, which he did with great difficulty. Fortunately, he asked to be accompanied by a union representative and was able to defend his promotion. A normal person would have reacted with compassion to the colleague's plight, but the boss saw this only as an opportunity to seize an advantage to try to put the boot in.

This example illustrates the way psychopaths operate. They are not limited by feelings for others and so can act with impunity to achieve their aims. As Cleckley (2015) writes in *The Psychopath: The Mask of Sanity*, addressing a psychopath, "You are not held back from any of your desires by guilt or shame, and you are never confronted by others for your cold-bloodedness. The ice water in your veins is so bizarre, so completely outside of their personal experiences that they seldom even guess at your condition." Should a normal person have a dispute with someone, they often may be prepared to go more than halfway in order to resolve the dispute. This does not work with a psychopath. If you give them an inch, they will take a mile. They regard this flexibility as a weakness and they will despise you for it.

The legal argument

A rationale for why psychopaths are so successful in conning their way through their careers is to do with what is termed the "legal argument." How this works is as follows. Because we assume that other human beings are trying to be "good" and honest, we do not usually question this assumption. When there is a dispute between two parties, we automatically think that the truth lies somewhere between the two extremes. What happens, however, in a dispute between an honest person who tells the truth and a dishonest person who always lies, is that the outcome does not match this reasoning. If the honest person is innocent, the only lie that he/she can tell is to falsely confess to a wrongdoing. In contrast, the liar can accuse the honest person of being at fault. The basic assumption that the truth lies between the testimonies of the two sides always gives the advantage to the one who lies, the psychopath. A good liar can twist the truth and can always make an innocent person look bad, especially if the innocent person is always honest and admits to sometimes making mistakes.

Psychopaths in science

Psychopaths sometimes reach high positions in scientific organizations, especially in recent times in which a managerial culture has flourished. This is not because they are outstanding scientists but because they are able to use their manipulative "skills" to exploit the system. In order to genuinely build a strong record in scientific research, one normally needs to put in the hard yards. This entails intensively following the relevant literature, designing and carrying out imaginative experiments to test hypotheses, and overcoming the inevitable disappointments and frustrations that are inherent in research. Psychopaths choose not to suffer these hardships. They may build a reputation based on some relatively trivial but topical research. They then embellish the research to impress a large number of their colleagues and especially their superiors. Once they achieve this and are promoted to relatively high positions in the hierarchy, it is then plain sailing for them. They do not need to continue to work at the bench. Instead of reading the literature, which is time consuming and unlikely to give quick dividends, they get their information by talking to their "lackeys" and by holding group discussions in which they pick the brains of colleagues to steal their ideas. They have no need to worry about initiating new scientific approaches. Their parasitic nature enables them to feed off their subordinates and, in any case, they insist on getting their names on a large number of publications as coauthors even though their contributions to the work may be insignificant.

What is the origin of psychopathic behavior?

There is some evidence that psychopathic behavior has a biological basis and has features of a disease. Some research has focused on brain imaging techniques such as positron emission tomography (PET) and magnetic resonance imaging (MRI). Two main theories have emerged from these studies. One suggests that the orbitofrontal region of the brain called the orbitofrontal cortex plays a role (Koenigs, 2012). It is known that this part of the brain is involved in conscious decision making. Another theory is that the amygdala is involved (Yang et al., 2009). This is a small almond-shaped structure that has a role in processing emotion and mediating fear. Since the orbitofrontal cortex, which does the "thinking," and the amygdala, which does the "feeling," appear to be interconnected, the two theories may not be mutually exclusive. However, the research in this field is ongoing and the question is far from being resolved.

Corporate psychopaths

In recent times, there has been an alarming increase in the demise of large corporations. This has resulted in employees losing their jobs and shareholders losing their investments as well as the capitalistic society losing some of its credibility. Incredibly, the senior directors of these failed corporations are frequently seen to walk away without taking any responsibility for the failures. They appear to not accept any blame and often move into other high-profile positions. Boddy (2011) has pointed out that these people present characteristics of psychopaths. In fact, he suggests that the recent global financial crisis (perhaps more accurately referred to as the Western world financial crisis) has its origins in the behavior of persons who are put into leadership positions in corporations but are really impostors who cause their destruction from within. These people have been called corporate psychopaths. They destroy the morale and emotional well-being of fellow employees (Hare, 1999). They do this by humiliating them, lying about them, bullying them, and blaming them for the mistakes made by the psychopaths. This can result in good people leaving and this undermining of human resources weakens the organization. Furthermore, it is thought that such people jeopardize the long-term success of an organization by doing whatever it takes to win contracts and by failing to meet their promises, thus damaging the reputation of the company (Boddy, 2011).

How do psychopaths attain high positions in corporations?

Ironically, the attributes that psychopaths possess (or pretend to possess) are similar to those that are sought for leaders in corporations. This link has been pointed out by Babiak and Hare (2006) and by Lipman (2013). Psychopaths can present themselves as highly successful by claiming the good work of others and by manipulation. Their innate cold-heartedness allows them to treat other employees harshly or to suddenly terminate their contracts without feeling any sense of remorse. These are sometimes seen as desirable attributes for corporate leadership. Some suggestions have been given by Lipman (2017) for preventing the hiring and promotion of psychopaths in organizations. These include having a sound internal succession program, as long times are needed to observe the true characters of individuals. In the case of hiring from outside the organization, real substantive accomplishments need to be verified rather than being persuaded by superficial charm and force of personality. Another important point is that, in an interview, there should be an attempt to glean as much as possible about the moral and ethical character of a candidate, something that is more difficult to achieve.

Coping with psychopathic behavior

Although the frequency of psychopaths in the workplace is quite low, their presence has a destructive influence out of all proportion to their numbers. It might seem that if there is a psychopath among a group and who has attained a position of power in the hierarchy, his or her effects could be countered if colleagues were to unite together. This sounds logical but it usually does not happen for the following reasons. First, hardly anyone understands what a psychopath is and therefore has no idea how to identify one. Second, psychopaths are very clever at hiding their traits. They usually only target one or a few victims and so their behavior does not appear abnormal to all the rest. Third, when someone is being unfairly treated, colleagues may see this as not being their problem and, especially in an environment that is highly competitive, are reluctant to interfere. As a result, most victims of psychopaths feel isolated and impotent to do anything about their predicament.

Aftermath: Surviving Psychopathy Foundation is a nonprofit organization that is a working coalition of psychopathy researchers, mental health professionals, victims, and family members of those affected by psychopathy. The purpose is to provide information and support to those whose lives have been placed at risk or negatively impacted by psychopathy. On its website (http://aftermath-surviving-psychopathy.org/), a moderated forum, a radio blog, and informational resources can be found.

How it happened

With the wisdom of hindsight, it is easy to understand how the unfortunate situation in CSIRO arose. It could and does happen to any scientific research organization. Once the initial decisions are made and the steps are put in place, the results are inevitable. It was decided in 1986 to contract a management consulting company to carry out a review of CSIRO. The effect of acceptance of the recommendations made by the company effectively was to introduce a management culture into a scientific organization. The decision was made by people who had little understanding of science. The ministry of the Australian federal government was concentrated in graduates from law with some from economics and commerce. Of the law graduates, one was the minister for industry and commerce and another was the minister for science.

It was stated by Rees (1995) that "managerial fundamentalism is apparent in its dogma, intolerance of critics, and gratitude for compliant staff" (p. 25). This is the exact opposite to what is inherent in the scientific method that was discussed in Chapter 2. Science rejects dogma and replaces it by critical thinking. Criticism is the lifeblood of science. It is recognized as being an essential component needed for the advancement of knowledge. Scientists do not seek compliance. They are only interested in arriving at the truth. How then could the imposition of managerialism on science be anything other than destructive?

The Executive of CSIRO, which had been comprised of outstanding research scientists, was replaced by a Board in which there was a minority of members with technical backgrounds. The Chief Executive in the new set up, who was an eminent scientist, was retained but, after his term had been completed, the appointments of subsequent Chief Executives were made with the requirement of management credentials in addition to scientific backgrounds. The culture changed inexorably from one in which outstanding research scientists enjoyed a certain independence, essential for creative work, to one in which they became subjected to a hierarchical system of management. Management became more important than science and this has been reflected in higher salaries for managers than for many of the active scientists.

The evolution of CSIRO from an organization with relatively autonomous researchers to a hierarchical management structure seems to have paved the way for impostors to enter and flourish, just as has been described earlier for failed corporations. Bullying and other antisocial behavior has been claimed on the Victims of CSIRO website. In a decision in relation to a compensation claim lodged by a former CSIRO employee, the Appeals Tribunal deputy president deemed the evidence of two CSIRO employees (one of them a senior executive) to be unreliable, providing no less than 128 false or misleading statements to the tribunal. Some of those who have been accused of misconduct have slipped out of the organization, in some cases leaving with full entitlements. All this behavior is similar to what has been reported by Boddy (2011, 2014) as a result of his observations on the failure of some corporations.

In response to the allegations of workplace bullying and other misconduct, CSIRO appointed a law firm (HWL Ebsworth) to investigate the claims. That meant that the misconduct was not investigated by an independent body but by a body appointed by the organization against which the complaints were being made. The investigation has produced reports in two stages. The report on stage 1, published in August 2013, found that there were problems at the CSIRO and "pockets" of particular concern but no toxic workplace culture of widespread bullying. This was despite 130 allegations being submitted. The report on stage 2 of the investigation was posted on the HWL Ebsworth web page toward the end of May 2014. It makes some recommendations for improvements in the organization to deal with the allegations of bullying. The Victims of CSIRO has posted a considered response to the stage 2 report. Up to a short time after the response, the group had received 100-percent negative feedback from its members and had pointed out many flaws in the investigative process. A detailed description of the investigation will not be given here and the interested reader is referred to the HWL Ebsworth and Victims of CSIRO websites.

Notes

1. www.brainyquote.com/quotes/authors/a/albert_szentgyorgyi.html

References

Babiak, P., and Hare, R.D. 2006. *Snakes in Suits: When Psychopaths Go to Work*. Harper Collins, New York.

Besser, L., and Phillips, N. 2013. Science second in toxic CSIRO work culture. *Sydney Morning Herald*, Technology section, April 19.

Boddy, C.R. 2011. *Corporate Psychopaths: Organizational Destroyers*. Palgrave Macmillan, UK.

Boddy, C.R. 2014. Corporate psychopaths, conflict, employee-affective well-being and counterproductive work behaviour. *Journal of Business Ethics* 121:107–121.

Cleckley, H. 2015. *The Mask of Sanity: An Attempt to Clarify Some Issues About the So-Called Psychopathic Personality*. Echo Point Books & Media, Brattleboro, VT.

Hare, R.D. 1999. *Without Conscience: The Disturbing World of the Psychopath Among Us*. Guilford Press, New York.

Hare, R.D. 2003. *Manual for the Revised Psychopathy Checklist*, 2nd ed. Multi-Health Systems, Toronto, Canada.

Koenigs, M. 2012. The role of the prefrontal cortex in psychopathy. *Reviews in Neurosciences* 23(3):253–265.
Lipman, V. 2013. The disturbing link between psychopathy and leadership. *Forbes*, April 25. www.forbes.com/sites/victorlipman/2013/04/25/the-disturbing-link-between-psychopathy-and-leadership.
Lipman, V. 2017. How to improve your odds of hiring the best people. *Forbes*, May 22. www.forbes.com/sites/victorlipman/2017/05/22/how-to-improve-your-odds-of-hiring-the-best-people.
Lopez de Victoria, S. 2008. How to spot a narcissist. *World of Psychology*, August 4.
Moses, A. 2010. CSIRO to reap "lazy billion" from world's biggest tech companies. *Sydney Morning Herald*, June 1.
Rees, S. 1995. The fraud and the fiction. In *The Human Costs of Managerialism: Advocating the Recovery of humanity*, S. Rees and G. Rodley, eds., 15–27. Pluto Press, Sydney, Australia.
Sheffield, H. 2015. The top ten jobs that attract the most psychopaths. *The Independent*, December 15.
Terman, L.M. 1955. Are scientists different? *Scientific American*, January.
Yang, Y., Raine, A., Narr, K.L., Colleti, P., and Toga, A.W. 2009. Localization of deformations within the amygdala in individuals with psychopathy. *Archives of General Psychiatry* 66(9):986–994.

chapter five

How the media influences public thinking

Democratic societies are blessed by having free media. Autocratic and theocratic societies do not have this luxury. Their media covers news but it is controlled in order to present it so that it is consistent with the ideology of the governing regime. A free media means that it is not subjected to oversight by government. It does not necessarily mean, however, that its readers, listeners, and viewers are presented with information that is always accurate and unbiased. Nor does it mean that the government, corporations, or lobby groups do not necessarily have an influence. Sometimes the information can be unintentionally inaccurate, while at other times it may be deliberately biased toward one political viewpoint. The media can have a huge influence on how the public get their facts and form their opinions. If a large proportion of the public is gullible, they may believe all the information they are fed. If much of this is false, many people will develop ways of thinking that are not based on truth. In order for members of a society to avoid this trap, it is imperative that they adopt a critical approach to everything they see, hear, and read. Scientific or critical thinking is essential if a democratic system is to prosper. We are going to look at a few examples of how the public is lured into false beliefs through the media.

The myth of the lemmings

Many people have come to believe that lemmings, small mouselike rodents, are capable of mass suicide by plunging into the sea. This belief is based on a 1958 film *White Wilderness* produced by the Disney Corporation, which showed the remarkable phenomenon of lemmings apparently voluntarily leaping to their death. The truth was that the filmmakers faked the scene by throwing the lemmings off a cliff (Cohen, 2014). This demonstrates how easy it is to instill a belief into a large number of people by a media organization. This was a relatively harmless deception, but its effectiveness raises the question of how easy it might be to convince large sections of the public of the veracity of more dangerous falsehoods.

The invasion of Iraq in 2003

In early 2003, a coalition of countries, led by the United States, invaded Iraq and removed its leader, Saddam Hussein, and the government. In order to gain the support of the people prior to the invasion, the U.S. administration embarked on a campaign to justify the decision to go to war. It was based on a presumption that Iraq possessed weapons of mass destruction (chemical, biological, and perhaps nuclear) and this posed a danger to the security of the United States and the world. This was not long after the attacks on New York and Washington in 2001, and these events were in the minds of citizens, making them sensitive to security concerns. The government used a series of strategies to persuade the people to support its decision to invade. These strategies have been outlined by Elliot D. Cohen (2014) and include, among others:

- Fearmongering (raising and lowering the alert level)
- Well-poisoning (calling people who opposed the war "un-American")
- Propagation of prejudice (media stereotypes of Arabs as terrorists and suicide bombers)
- Jingoistic appeals (positioning the American flag behind news anchors on Fox News)

In a January 2003 CBS poll, 64 percent of Americans approved military action against Iraq. Now that more than a decade has passed since the Iraq invasion, it is easier to analyze the events. No weapons of mass destruction have been found. Thus, the main reason given for the invasion appears to be false. There is evidence that the U.S. administration had made the decision to attack Iraq well before 2003. Furthermore, a copy of a leaked memo obtained by British journalist Michael Smith and published in the *Sunday Times (London)* seems to suggest that the U.S. administration, prior to the invasion, did not believe that Iraq had weapons of mass destruction (Smith, 2005).

The purpose here is not to carry out a detailed analysis of the political decision making but to show how a government, aided by the media, can manipulate the beliefs and opinions of the people. As Cohen (2014) writes: "We Americans are not helpless victims of the politico-corporate media establishment. Victims yes, helpless no. We largely permit ourselves to be duped and manipulated. If you think otherwise, then you are subscribing to a view that makes lemmings of us all."

The Chilean military coup of 1973

The role of the media in influencing opinion is not restricted to one side of the political spectrum. In the 1970 Chilean election, a Marxist president, Salvador Allende, was elected to lead a coalition of socialist-oriented parties. In 1973, the elected government was overthrown by a military coup led by army general Augusto Pinochet. Understanding of these and subsequent events has been considerably influenced by the media. Many people around the world hold the belief that Pinochet was a ruthless dictator who removed a democratically elected government and inflicted torture on his people. A juggernaut campaign effected by the media had a large part to play. The true situation is not so simple if one adopts a critical approach to examine the events. I was living in Chile at the time of the election of Allende and so may be able to provide a less partisan viewpoint of the happenings than many in the media who were not firsthand observers.

The first thing to note is that there were three candidates in the 1970 presidential election. Allende secured 36.62% of the votes; the right-wing candidate, Jorge Alesandri, 35.27%; and the third candidate, Radomiro Tomic, 28.11%. If no candidate receives over 50% of the votes, the Chilean constitution decrees that both houses of the National Congress (not the people) must vote on the two candidates who received the most votes. Allende's presidency was eventually ratified (Allende 78.48%, Alesandri 17.95%.) after he agreed to sign a "Statute of Constitutional Guarantee," promising not to undermine the Chilean constitution.

After the election, Allende attempted to implement his agenda. The agenda included nationalization of industries, such as the large copper mining companies and raising the wages of workers. He met with much opposition, as he did not have a majority in the senate. The government was squeezed from both sides of politics: the conservatives on one side and the extreme left on the other, who were impatient for socialist reforms to be put in place. A group of young men, self-appointed "Amigos del Presidente," formed a type of security guard that traveled everywhere with the president to guarantee his protection. There were an unauthorized series of farm and factory seizures by radical supporters of Allende's coalition, who felt empowered by the new regime. An acquaintance of mine had bought a bakery a short time before the election and it was progressing well. It was taken over by a group of regime supporters who did not have the expertise to run the business. As a result, the machinery failed, the production ceased, and the bakery was left abandoned. Similar results occurred with the takeovers of farms. Many of those who took over had no experience of farming and some seemed more concerned about imposing their ideology than running an enterprise for profit. A large section of the more conservative population was appalled at the apparent trend toward anarchy of a country that had a history of peace and political stability for many decades.

During this time, there was an exodus of business people from the country and a crippling strike of truck drivers was staged. All this helped to contribute to a serious crisis in the economy. A view formed among a large proportion of the people that the country could not continue along the same path and there was a strong movement for the military to take control. The 1970–1973 period was one that created a sharp polarization of the Chilean people. Those who supported the Allende government naturally felt cheated that, having been democratically elected, it had been removed by a military coup. Thus, there was opposition to the military rule, which, in certain cases, led to insurgence and guerrilla tactics. Regardless of what one thinks about an elected government being overthrown, the reality was that it happened. The military was now in charge and its responsibility was to act in the national interest. This it did, as military regimes usually do, with ruthlessness and effectiveness.

The purpose in summarizing the events in Chile is to try to present a more balanced appraisal than the one that is currently held by many people around the world. The generally accepted view is that a democratic government was overthrown and replaced by a cruel dictatorship that engaged in torture of civilians. This simplistic view has been largely imposed by the media. It has also been reinforced by those people who fled into exile after the military coup, settled mainly in countries with socialist sympathies, and influenced how their host countries and the world regarded the events. Many facts are overlooked. For example, it is not generally admitted that, at the time of the coup, the economic situation was dire and that in subsequent years, the national debt was repaid. Chile rose to have become the star performer among South American economies, which can to a large extent be attributed to policies initiated by the military regime.

How can we decide what is the truth?

This is not to claim that the summary I have given is the truth. Truth about contentious issues is never easy (and, in some cases, perhaps impossible) to establish and usually cannot be described by a simple statement. Perception of truth will be different, depending on whether the person was one whose property was confiscated during the Allende rule or whether it was someone who lost a relative as a result of the purge following the coup. Scientific thinking never claims to arrive at an absolute truth. The best it can do is to discover the facts as much as is possible and to examine them rationally without allowing emotion to intercede. What can be said with some certainty is that it is possible to influence public opinion by a media that presents a simple message and continuously repeats it. A large proportion of people do not have the time or the inclination to delve into an issue. They like to have a simple message and, once they have that, they will form an opinion and no amount of logical argument can easily cause them to change. Although one is never justified in claiming to know the truth, it is possible to know when one analysis of events is better than another. If one argument is based on a more detailed examination of the facts than another, it is likely to be closer to the truth. Another criterion is that, if an argument is based on a consideration of all sides, it will usually be closer to the truth than one that approaches an issue from only one side. In the case of complex issues, we should accept that arriving at an absolute truth is often not feasible and we need to be content with relative truth.

Contrast between submissions to scientific journals and to the mainstream press

The process for submitting papers to scientific journals is roughly as follows. The submission is sent to the editor who makes a decision on whether it should be sent out for review or rejected without further review. A submitted paper may be rejected by the editor if it is considered inappropriate for that journal, whether no significant new knowledge is presented, or whether the science is not sound. The author is then informed of the decision and the reason(s) for rejection. If it is not rejected, the editor will send it to a number of reviewers (at least two) who will, after examining the paper, forward their reports to the editor. The reviewers will either recommend rejection or tentative acceptance with the proviso that certain criticisms that are raised need to be addressed. The editor then makes a decision based on the reviewers' reports and his/her own judgment and informs the author. The author may then have to revise the manuscript in response to the comments until the editor (and reviewers) are satisfied that the revised version meets the required standard. The evaluation system thus goes a long way toward ensuring that submissions are fairly considered.

Consideration of submissions to scientific journals is thus characterized by transparency. In addition, it gives the authors feedback that can help them to make improvements or, if the paper is rejected, to provide information that can help them in future submissions. In the case of submissions to newspapers or other commercial publications, the procedure is usually quite different. Those who submit letters to the editor or submit book proposals to publishers, for example, are often not acknowledged. Their contributions may be published but, if not, they are not given reasons why they were rejected. Thus, unlike submissions to scientific journals, there is no feedback given to the authors that could assist them in preparing contributions that might be accepted. As one example that appears to be typical, a prominent Australian publisher, Allen & Unwin, invites book proposals in a system called Friday Pitch. Authors are advised that if their proposal is not accepted, no further communication will be entered into. Thus, authors are left in limbo without knowing whether their submission will be considered, even though the courtesy of a simple e-mail could have saved them from an anxious long wait. Such an arrogant response by publishers seems to be rather unique among those who depend on the public for their business. The process for reviewing articles or books submitted to the commercial press therefore does not have the quality control that scientific journals demand. This, despite the fact that their staff members are remunerated, whereas the scientists who carry out the onerous reviewing tasks mainly do this voluntarily.

Issues that seem to be ignored by the mainstream media

A comment that has been bandied about is that those who do not read newspapers are uninformed, and those who do read newspapers are misinformed. This seems to be rather unfair but, like many sayings, there may be some truth in it. Most media outlets present news and opinions that are slanted toward certain political viewpoints, although it must be said that the more reputable ones make great effort to arrive at a balance, catering for all sides of politics. Those who do not read newspapers may be uninformed, but those who do read them can also be uninformed simply because some issues that should be seen as important are ignored. Those items of news that fall through the cracks may sometimes be taken up by alternative media such as talk-back radio. Talk-back radio is a relatively new form of media communication. It gives the opportunity for members of the general public to express views that they would not normally be able to.

The main aim of a media corporation is to maximize its profits. This therefore sets limits to the balance of news items that are presented. Issues in which all sides of the political spectrum are in agreement are likely to be less newsworthy than ones that are divisive. As a result, some items of news are not given high emphasis even though they should normally be of great interest to the general public. We will look at only two of the many examples of news items that have been seemingly neglected by the mainstream media.

Agenda 21 or sustainable development

Agenda 21 (the "21" referring to the twenty-first century) was a product of the United Nations Conference on Environment and Development (UNCED) held in Rio de Janeiro, Brazil, in 1992. Its aims were described in a 300-page document that had been developed over several years prior to the Rio conference. It was then agreed to by 178 governments. The key objectives were global and were directed to the following:

1. Combating poverty, especially in developing countries by changing consumption patterns, promoting health, and achieving a more sustainable population.
2. Protection of the environment by combating deforestation, conserving biological diversity, and controlling pollution.
3. Strengthening the role of special groups, including youth, women, local authorities, and indigenous peoples.
4. Implementing the aims by using science, technology transfer, education, and financial mechanisms.

It is not, at first thought, possible to see an objection to these aims. Preservation of the world's finite resources for future generations seems to be a noble objective and this seemed to be the view of the 178 signatories at the Rio conference. The agenda is based on the concept that current lifestyles and consumption patterns, particularly in the more affluent populations of developed countries, are not sustainable. To meet the resulting challenges requires some sacrifice of national sovereignty and its replacement by a global program. The development of this program under the auspices of the UN will require public participation and involvement of all levels of government. It is envisaged that it would be driven from a grassroots level in which local councils would play an important role.

Although Agenda 21 has been promoted as the salvation of the planet and has been supported by opposing political parties in many countries, there have been those who see it as an insidious plot for world domination and control of individual's lives and liberties. Its true objectives are asserted to include "restructure of the family unit, which means basically the state will take care of your children, with a keen eye toward indoctrinating them into state control over family allegiance, abolition of private property with citizens crammed into 'stack and pack' hi-rise apartment buildings, close to railroad tracks while cars will not be allowed" (Femine, 2013). The opposition to Agenda 21 is typified by the U.S. state of Alabama, which passed a law banning any and all sustainability projects of the UN from being enforced in the state.

British immigration since early 1980s

Since the early 1980s, Britain has undergone a spectacular demographic change. The country was largely unidiverse prior to then. For example, before the Second World War, 70 to 75 percent of British DNA had been

British for 13,000 years with later migrations having little impact, including the Anglo-Saxon invasion, which contributed about 4 percent of British DNA. As a result of changes in immigration policies, a quarter of the births in England in 2009 were to foreign mothers. In 2009, a former Labour party speechwriter, Andrew Neather, stated that the dramatic increase in immigration in the previous years had been politically driven (Whitehead, 2009). Although Neather subsequently backtracked, release of documents under the Freedom of Information Act suggested that he had been telling the truth. These documents revealed that the policy of the then-Labour government was "a deliberate attempt to change the face of British society" (Slack, 2010). The relevant document showed that the original document had included mention of the government's social objectives for migration policy. These references had, however, been edited out, apparently due to concerns about how the public would react to attempts to use immigration policy to change the ethnic mix of the society.

The aforementioned two topics are examples of issues that have been largely ignored by the mainstream media, at least in Australia, but would normally be expected to be of vital concern to the public. One side of the argument presents Agenda 21 as a benign plan to save the planet from destructive forces. The counterargument suggests that it is a dangerous agenda with the aim of subjecting people to domination by a world government that is intent on erasing the sovereignty of countries. The other topic, the use of immigration policy to change the cultural makeup of a nation, would also be expected to resonate with the public. Currently, there are large masses of migration entering various countries and, in some cases, national governments give the impression of not taking strong steps to prevent it. An ulterior motive for encouraging immigration from specific ethnic cultures can be to increase the quantity of votes for the party currently in power. The failure of the mainstream media to cover such issues as these leaves some unanswered questions. Fortunately, these issues have been raised by talk-back radio programs such as those of the Macquarie Radio Network in Sydney in which broadcasters Brian Wilshire and Michael McLaren have been in the forefront. The audience that talk-back radio commands is, however, quite small compared to that of print media and television.

The opinion writers

Newspapers employ columnists who regularly contribute with opinion articles. Some opinion writers are outstanding journalists who strive to present balanced viewpoints, whereas others inject their political bias into their writings. Since the latter are permanently employed as contributors, they can exert a large influence on a section of the readers who may not apply the critical thinking that is needed to properly evaluate the writings.

Many people choose to read articles or listen to programs that are only in accord with the opinions they have already formed. They thus do not allow their minds to be challenged by viewpoints that are at odds with those strongly held opinions. I have observed that many who form opinions about political figures or commentators, do not base their opinions on reading or hearing the original sources. Instead, they depend on second-hand information, which is likely to be biased.

What are the consequences when an opinion writer is shown to be wrong? A problem faced by Australia, particularly in the period 2007 to 2013, was the influx of illegal arrivals by boat. This was fanned by a well-organized people-smuggling business and by a government that seemed impotent to stop it. During this period, some 800 boats arrived, bringing about 50,000 asylum seekers, while it is estimated that about 1200 perished in the attempt. Apart from the tragic loss of life, another obvious result has been a huge financial cost to the nation as well as posing a threat to security. At the federal election in 2013, the Liberal-National coalition, which was elected, gave an undertaking that it would stop this people trade and restore sovereignty to the borders. One of the stated policies was that the boats would be turned back if safe to do so. In an article in *The Australian* newspaper, Chris Kenny (2014) exposed how a good number of the elite opinion writers were adamant that this could not be accomplished. In fact, as it turned out, the policies put in place by the government that was elected in 2013 have proved successful in stopping the boat arrivals. There do not seem to have been much in the way of admissions by these same opinion writers that they had been wrong. In fact, Kenny quotes one of them, who had maintained for three years that the boats could not be turned back, as tweeting, after it became clear that the policy had been successful, "Boat turn backs was always going to work. But at what cost?" This typifies how some people, not only opinion writers, often evade admitting their erroneous views by trying to deflect the original issue into a new one. The unfortunate consequence is that many who have not learned to think critically will swallow the deflection and not recognize it as a dishonest strategy. Some opinion writers have it very easy. They can relentlessly attack politicians and, as we have seen, can be wrong without being adequately held to account. Politicians, on the other hand, trying hard to do their job as best they can, need to be continuously on their guard to avoid a "gotcha" attack by the ever-vigilant opinion writers and broadcasters, some of whom wait like hawks, ready to pounce. Politicians, as a result of the 24-hour coverage of events, are closely monitored in what they say. The opinion writers, by contrast, appear to have a considerable amount of liberty. Should they be criticized by politicians or the public, they may respond to the criticism by calling it partisan and justify their actions by appeal to the freedom of the media. Freedom of the media is precious but, like all things, it can be abused.

How a democratic system can break down

Winston Churchill implied that democracy is the best form of government. It is a fair system in that it gives voters in the electorate equal rights. The candidates are then elected on the basis that a majority of the voting population believes that they are the best persons to govern the country. In theory, this should work well, that is, providing the electors are well informed about the issues and are able to sensibly evaluate them. But what if the voting public is not well informed or is not capable of rationally evaluating the issues. The inevitable result is that the nation's best interests may not be served by people who are elected. The potential weakness of a democracy is that the vote of a person who is well informed and critically examines the issues carries the same weight as a person who is ignorant and easily indoctrinated.

Electorates are made up of a mixture of various groupings. A proportion is made up of critical thinkers. Another proportion is made up of people who are only able to think in a certain way. Their ears are tuned only to what they like to hear. Another group is not prepared to give very much thought to the political process and can be easily won over by slogans. Pervading the electorate is the consideration of what would be in the best interests of the individuals who are deciding how to cast their votes. Political candidates must therefore organize their campaigns based on a consideration of the makeup of the electorate. This is a makeup that may have remained stable in the past or may be rapidly changing.

Political analysts and opinion writers, in trying to explain polls and election results, tend to focus on politicians and where their policies may be going wrong. There is little examination of the composition of the electorate. What proportion of the voters are critical thinkers and well informed, what proportion are easily swayed by rhetoric, what proportion form opinions based on what they read on social media? Politicians need to have this information in order to organize their campaign strategies. For example, if it is apparent that there is only a small proportion of critical thinkers, there is little mileage in directing the campaign to rational arguments about the issues. We need to remember that the priority of politicians is to get elected. If there is a large section of the electorate who are not interested in examining the issues and do not become well informed, the most effective strategy for some politicians is to continually attack a person on the opposing side, preferably someone near to or at the top. If they can convince these people that this is a bad person who does bad things and so should be hated, then they have their votes.

The rise of social media

A change that has occurred in recent times has been the spectacular increase in the use of social media, which has been taken up especially by the younger section of the community. The great difference between the social media and the traditional media is the almost complete absence of quality control of the social media. Anyone can place their opinion on record without the strict filtering that is in place for most traditional media outlets. Many people are now getting all their information and forming opinions based on what they read on Internet sites such as Twitter and Facebook. People who are unwilling to read serious articles by qualified writers find it easier to follow Internet posts that do not require them to depart too far from their comfort zones. The posts that they choose to read tend to be those that accord with their preconceived views. The social networks therefore act as echo chambers in which everyone agrees with everyone else. The absence of provocative discussion inevitably leads to a degeneration of critical thought.

How to think for yourself

A democratic system can only be truly successful if a large proportion of the electorate has the capacity for critical thinking. Cohen (2014), although admitting that a single article is not sufficient to cover all the rational thought processes that are needed to promote democracies, suggests six practices that are crucial to thinking for ourselves. These can be succinctly summarized as follows:

1. Ask for explanations. News items are usually presented as a series of unconnected events with no attempt to explain them. It is up to us to dig deeper to gain understanding of what is really behind the events. This means gathering evidence from multiple sources and not just from one corporation that may well have its own agenda. Explanations need to be based, not on speculations, but on facts.
2. Look for consistency. Facts need to be consistent. If an explanation is inconsistent with the facts, then it is not probable. It is necessary to check several independent sources before accepting.
3. Question the status quo, don't just believe it. It is important not to believe something because it is a popularly held belief. The myth of the lemmings is a good example of this.

4. Believe only credible authorities. Since none of us are experts in all areas, we need to rely on the testimony of others who are experts in their field. However, it is preferable to rely on several independent authorities rather than just one.
5. Watch out for fearmongering and demagoguery. Fearmongering works by exaggerating the consequences of something unpleasant happening. Some pundits who exaggerate the future effects of climate change fall into this group.
6. Beware of stereotypes. Stereotypes rely on inadequate evidence. Children often form mind-sets as a result of what they are taught through socialization or images presented by the media. To prevent ourselves from adopting these mind-sets requires concerted efforts to resist their acceptance and to realize that all human beings should be considered as individuals.

Cohen urges people not to be puppets of the politico-corporate media establishment. He suggests that everyone should become an investigative journalist, using criteria such as the six described. Failure to do so means that our freedoms will inevitably be threatened. In order to counter the false-to-fact claims that we are subjected to, Cohen says that we should fight back by "arming ourselves with the most powerful weapon known to humankind: rational thinking."

References

Cohen, E.D. 2014. Digging deeper: Politico-corporate media manipulation, critical thinking, and democracy. *Project Censored*, chap. 5.

Femine, L. 2013. Agenda 21 revealed—You need to know this. *Las Vegas Guardian Express*, September 21.

Kenny, C. 2014. Labor in denial over the obvious success of Abbott's asylum boat policy. *The Australian*, November 1.

Slack, J. 2010. How Labour threw open doors to mass migration in secret plot to make a multicultural UK. *Daily Mail*, February 11.

Smith, M. 2005. The Downing Street memos. *Sunday Times of London*, June 16.

Whitehead, T. 2009. Labour wanted mass migration to make UK more multicultural, says former advisor. *The Telegraph*, October 23.

chapter six

Dangers to progress in science

The great philosopher of science Karl Popper (see Chapter 2) emphasized that science has a need to grow. Progress in science does not mean the continuous accumulation of observations, but it is more to do with the examination of theories and, by criticisms, their replacement by better ones. Thus, science grows by increasing our understanding as a result of conceptual advances. If this does not happen, science stagnates. Science is unique in that, by its nature, errors should be systematically criticized and, in time, corrected. This leads to progress. In most other human endeavors, there is change but not necessarily progress. We are going to look at some of the obstacles that stand in the way of progress in science. Several sources of danger to progress, suggested by Popper, were mentioned in Chapter 2. We will look at these in greater detail here.

Dangers to progress of science suggested by Popper

The first danger proposed by Popper was a lack of imagination or real interest. Intellectual curiosity has waxed and waned throughout history. There have been periods when it has been low or practically nonexistent. Such periods have sometimes been followed by periods of intense intellectual curiosity such as during the period that has been termed the Enlightenment. This was a period stretching from roughly the middle of the seventeenth century through the eighteenth century. It was a time when Western thought and culture underwent a revolution. It involved a release from superstition, prejudice, and dogma, and the use of one's own reasoning to acquire understanding. The conditions needed to enable this to occur require a certain quality of life as well as a certain amount of political freedom. This creates the conditions for humans to question traditional acceptance of beliefs. It led to a revolution in science that has continued more or less to this day.

Great scientific advances were made during the Enlightenment and these have gathered momentum throughout the nineteenth and twentieth centuries. Science has created technologies that have enhanced the quality of human life. It has done this by revolutionizing, among other things, transport, communication, agriculture, and medicine. The period from the 1950s to the early 1970s has been called the Golden Age. This was because global economic growth was especially high. This period also coincided with great scientific and technological advances—semiconductors, lasers, nuclear power generation, computers, and plastics to name a few. It was also the period when spectacular events were enacted such as sending satellites into earth orbit and traveling to the moon. These events helped to motivate young people to take up science as a career. It may be that this period was a pinnacle and we may now be seeing a decline of interest in science. This is reflected in falling enrollments in science courses and the lower numbers of high-quality students entering science, as has been noted in some countries.

Development of the Internet in recent times has facilitated access to information and allows scientists, as well as the general public, to quickly and conveniently retrieve available information. This is proving to be a great asset for scientists. There are also accompanying dangers. Publish or perish is a cliché that is often applied to describe how the performance of individual research scientists is judged. The rate at which a scientist publishes papers is used as a measure of his/her scientific output. The pressure to publish in order to enhance promotion prospects, to support applications for positions, or to secure grants, acts as an encouragement for researchers to carry out superficial investigations. Scientific research should be a relentless pursuit of the truth. Those who carry out honest research with the aspiration to discover genuine new knowledge usually spend a fair amount of time traveling up dead-end roads. When Albert Einstein began at Princeton University, he was asked what he would require for his study. He replied, "A desk, some pads, a pencil and a large wastebasket to hold all of my mistakes." In today's highly competitive world, some scientists may prefer to cut corners and may not be prepared to spend time on dead-end roads. An essential component of good research, however, is making mistakes and learning from them.

Obsession with number of publications

Frequently, the output of a scientist is assessed by the number of publications. Performance assessments are often made by people who are not closely familiar with the research field. They may then rely on the number of publications without being able to evaluate the quality of the work. Most of the great advances in science have been made by those who have had the freedom to direct their whole attention to discover real knowledge. Some of those who have made the greatest contributions may not have set out to produce a large number of publications, but the papers they published have turned out to be beacons that influenced the thinking of those who have followed.

The traditional role of scientific publication has been to record new results and concepts and to keep other researchers informed of recent developments. In earlier times, it was relatively easy to keep abreast of advances in a field. There were comparatively fewer hard copy journals and books that were available in order to keep well informed of progress. Although the rate of growth of journals and, as a result, scientific publications, has been increasing continuously for some time, the current sheer volume is creating problems. The space required for storing journals and the associated costs are causing libraries to cancel subscriptions. The introduction of electronic publishing has alleviated the problem to some extent. It has been reported that there are approximately 28,100 active peer-reviewed scholarly scientific journals (Boon, 2016). The annual rate of growth of scientific papers is presently estimated to be in a range of 5 to 15 percent, the rate varying between different scientific fields and between different countries. It is estimated that the number of papers published is doubling about every nine years. Does this mean that scientific knowledge is increasing at this rate? I think it would be hard to find anyone who would want to answer this question in the affirmative. If the amount of new knowledge is not proportional to the number of published papers, how then can we rationalize this explosion in published work? There seems to be two main answers that are interconnected. First, much of the scientific literature is composed of papers that do little more than record observations. As pointed out by Popper, progress in science does not mean the accumulation of observations but is more about advancing our understanding by examining and improving theoretical concepts. Second, although it is sad to admit, a primary incentive for publishing papers is often not to advance real knowledge but to enhance reputations and to improve promotion prospects. Let us look at two examples that illustrate these points.

Quantity and quality in scientific publications

In 1908, Albert Einstein published three papers on different topics, each of which was a milestone in theoretical understanding. These dealt with the topics of special relativity, Brownian motion, and the photoelectric effect. Special relativity revolutionized thinking about the physical universe and, as a spinoff, led to an understanding of nuclear reactions. Brownian motion increased understanding of molecular behavior and the photoelectric effect has been the foundation for modern electronics. Einstein was working at the Swiss patent office carrying out rather mundane work as a patent examiner. The work did not demand great intellectual effort. This allowed him to devote his time to pondering problems of the universe that interested him. He did not need to produce a lot of publications to further his career. His motivation was an intense curiosity to understand phenomena that other scientists had observed but could not fully explain. The result was that he introduced concepts that had not been dreamed of previously and which would change the direction of subsequent thinking.

In contrast to the three contributions of Einstein, I happened to recently read an article that mentioned a scientist who was credited with having published more than 800 papers in peer-reviewed journals. This number of papers was presented as a pointer to the eminence of this scientist. It is fairly common for scientists to have published large numbers of papers, in the many hundreds, these days. Scientific literature is seeing an ever-increasing number of multiauthor papers. It is sometimes difficult to ascertain what contribution each author makes, but it is an effective way for expanding the number of publications of each individual author. For those who amass such large numbers of publications, we are justified in looking beyond the number and to asking pertinent questions about the work. Publication of large numbers of papers is often achieved by those who hold positions in which they supervise many research students and technical assistants. The research may in some cases involve a series of similar experiments, each being slight variations of each other but sufficiently different to merit separate publications. If the experimental work could be distilled into one or two papers that summarize a conceptual advance, this might significantly alter the thinking and direction of research of those who follow. It would also make the task of reading the literature for other researchers a good deal easier and would save paper. This, however, is not always the case. The scientific literature is often cluttered with papers that simply describe observations without advancing theoretical understanding. This does not equate to scientific progress, as pointed out by Popper, and can lead to stagnation of science.

There are more problems related to the explosion in published papers. Based on a 2008 study, Evans (2008) reported that as more articles appeared online, scientists cited fewer of them in total and cited more recent ones with higher frequency, suggesting that older literature was no longer being read and/or cited. Much of the older work was still in hardcopy format. This illustrates the dangers of relying only on keyword searches to become informed about previous work, a procedure used currently by many researchers.

Citations and impact factors for measuring merit

It is becoming accepted that number of publications is not the most reliable way to assess a scientist's performance for determining promotion or for awarding research grants. Quality is a better guide than quantity. This, however, is not so easy to measure. One way of quantifying quality that has been used is to measure the number of times papers are cited by other authors. It seems reasonable to expect that the more value a paper has, the more it will be referenced by peers. A paper that presents new concepts that influence the direction of the work of subsequent researchers is likely to be highly cited. In contrast, a paper that seems to contribute nothing of value may not be cited at all. Thus, citations have been adopted as a measure of research excellence. Scientific journals are being assigned impact factors. The impact factor essentially calculates the ratio of the number of citations to papers published in the journal to the total number of published papers over a set period, say one year. Impact factors for individuals or for institutes are calculated in a similar manner. The criterion of citations for assessing quality of scientific research is not free of criticism but, at least, it seems to be a better measure than the number of publications. Some of the problems with using citation rate as a measure of scientific excellence have been succinctly pointed out by Squires (1992). Review articles or articles describing methodology are more likely to be cited than others. If an article is found to have a weakness, it may also be frequently cited. The impact factor for an institute can be raised because of one or two highly cited papers. If this institute produces a relatively small number of publications, the effect of these papers will have a greater contribution to the impact factor than it would for an institute that publishes a much larger number of papers. Another consideration is that reports of highly specialized research may make a substantial contribution to scientific knowledge but may not be cited often because there are few researchers actively working in that area.

Lack of citations to publications

In a study of the top 10 percent of all scientific journals worldwide from 1981 to 1984 by the ISI, the Institute for Scientific Information (Hamilton, 1990), it was revealed that 55 percent of papers were not cited in the first five years following their publication. An earlier study of articles in the hard sciences by the ISI between 1969 and 1981 found that only 42 percent received more than one citation. Assuming that a similar trend applied to papers in the 1981 to 1984 study, then it might be concluded that as many as 80 percent of papers published during that period were never cited more than once. It also needs to be remembered that self-citations, in which authors cite their own work, account for 5 to 20 percent of all citations. It should be noted that the evidence for lack of citations for scientific articles has been challenged by later empirical data (Lariviere et al., 2009).

An excess of poor-quality research

Apart from the controversy in regard to citations as a measure of scientific value, there is a widespread belief that there is an excess of poor-quality research currently being carried out and published (Bauerlein et al., 2010). The main cause for the steep increase in published scientific articles is attributed to growth in the number of researchers. Scientific research can be compared with other creative activities. We do not need more composers to compose more music. What is needed is to compose great music. Similarly, we do not need more scientists. We need only more scientists who produce good science. Expansion of the research ranks has been put forward previously as a reason for the decline in scientific quality (P. Stephan, as cited by Holden, 1991). Stephan claimed that the increase in the number of PhDs has bred an unhealthy competitive atmosphere that has created a profession that is overcrowded with risk avoiders. These people were more worried about the next grants than about intellectual creativity. It seems ironic that this should have been pointed out by an economist against strong objections from notable scientists who were calling for more PhDs.

Negative flow-on effects from mediocre research

If there is an excessive number of scientists producing mediocre research, this has a snowballing effect. There will be a lesser proportion of elite scientists to act as reliable reviewers for journal articles, inevitably lowering the standard needed for critical evaluation of submissions. Those who are regarded as competent reviewers are overburdened with requests to review. The reviewing load is then shifted to inexperienced reviewers who may not have the confidence or the courage to recommend rejection of manuscripts. Conversely, a lack of competence can lead to rejection of manuscripts that have merit. A classical example of this was the rejection of Edward Jenner's report of the first vaccination against smallpox (Behbehani, 1983). The intense competition among researchers, noted by Stephan, causes them to be so concentrated on preparing articles for publication that they do not assign the time needed to thoroughly peruse the literature. As a result, much good research that was reported at earlier times is ignored. This can lead to good research being repeated, but the repetition may not be done as soundly as the original. Although it should be recognized that there are pockets where high-quality research is being performed, it is apparent that a large proportion of scientific output is pedestrian. This is filling the literature with mediocre publications that are not contributing to exciting scientific advances. In the first paragraph of this chapter, it was proposed that science is unique in that, by its nature, errors are criticized and, in time, should be corrected. This is the ideal situation. If, however, quality research becomes buried in what Bauerlein et al. (2010) describe as an avalanche of low-quality research, there is a danger that this assertion may not apply. If the research activity is carried along by researchers who choose not to carry out the due diligence that is required for adequate scrutiny of previous work, it is possible for good-quality research to be lost, possibly forever.

My own experience gives support to this. At the beginning of my scientific career, I was carrying out research in a rather specialized field. At this time, some papers were published that appeared to me to provide a solid foundation for the research field to progress. When, after many years, I recently spent some time catching up with current literature in the field, I was disappointed to realize that this sound research had not been followed up. Instead, the field seemed to have gone backward. There seemed to be a plethora of inconsequential papers. Science is supposed to advance by researchers building on what earlier workers had contributed. This did not seem to have happened in this case, which caused me to question why. The impression I was left with from my reading was that the current researchers had not either acquired the fundamental knowledge of basic science or that they lacked the imagination that the authors of the earlier papers had possessed. I suspected that they were not motivated by the same passionate desire to discover new knowledge but may have been more focused on accumulating publications. This brings us back to the warning made by Popper that lack of imagination and real interest can be an obstacle to scientific progress. The focus on publishing papers is paralyzing the spirit of inquiry, and imagination that is essential for science to prosper. As remarked by Einstein, "Imagination is more important than knowledge. For knowledge is limited to all we now know and understand, while imagination embraces the entire world and all there ever will be to know and understand."

Not more scientists but more good scientists

There seems to be justification for rejecting the idea that more scientists are necessarily needed. Only good scientists are needed. These are people who have a genuine passion for discovering knowledge and are prepared to accept the sacrifices that this requires. In today's world, it is not easy to identify the more gifted scientists, partly because of the emphasis that has been placed on number of publications as a measure of scientific merit. Bauerlein et al. (2010) have offered some suggestions for correcting the problem of overpublication. One suggestion is that, instead of submitting a complete list of publications for those presenting their credentials to support promotion cases or applications for grants, the number of publications should be limited to three, four, or five. This would not only make it easier for reviewers to evaluate performance but would encourage candidates to focus their research more on quality. Other suggestions are to set a limit on the number of pages in a submitted manuscript, as is now done by some journals, and to give more attention to citations and impact factors for assessing quality of papers.

Effects of external and internal forces

At present, a large amount of public money is being spent on science. It is therefore important that the public maintains its confidence in the integrity of the scientific enterprise. Should this not be the case, the result will inevitably be to increase regulation and oversight of science. This is what has happened to some scientific organizations, resulting in the imposition of managerialism, as has been described in Chapters 3 and 4. It results in a vicious circle. Lack of confidence in science leads to greater control, which, in turn, causes restriction of freedom of scientists and this inevitably leads to a lowering in quality of science. The research environment changes as a consequence of external and internal forces, as has been pointed out by Alberts and Shine (1994). The external forces have their origin in greater demand for accountability. The internal forces arise from the intensified competition for research positions and funding. This competition puts stress on researchers and, in certain cases, can promote undesirable behavior and dangerous shortcuts.

Misconduct in science

The aim of science is to search for truth. Scientists are therefore expected to conform to strict standards of honesty. There are, however, abundant opportunities that can tempt scientists to deviate from these standards. Misconduct in science takes many forms. It covers a wide range of behavior from blatant fraud, through altering experimental data, plagiarism, and failure to give appropriate credit to the work of others. Fraudulent practices may be the most easily identified form of misconduct, although it may, in certain cases, take some time to uncover. One famous example is what has been referred to as Piltdown Man. In 1912, fragments of a skull and jawbone were recovered from a pit at Piltdown in West Essex, England. This find led to controversy for several decades. It was suggested that the fossil may have been the "missing link" in the evolution of humankind. At that time, a continuous transition from the early primates to humans in the evolutionary path had not been clearly identified and it was proposed by some that this could be the link that hitherto had not been discovered. Although it was revealed in 1923 by Franz Weidenreich, an anatomist, that the bones were a composite of a human cranium and an orangutan jawbone, the debate continued. Then, in 1953, the newspaper *The Times* published evidence gathered by many experts that Piltdown Man was a forgery. In the time between the initial "discovery" and final exposure of the fraud, many scientists' time was wasted and science was temporarily put off course.

More subtle forms of unethical behavior

The rather obvious examples of misconduct mentioned above such as data alteration, plagiarism and fraud are relatively easy to identify. In some cases, this has led to prosecution and accompanying disgrace for the transgressors. There is a general belief that these examples are relatively few. However, if the subject is probed more deeply, it has been suggested that unethical behavior is not so clear-cut, nor is it so rare (Martin, 1992). The cases of misconduct that have been exposed represent the tip of an iceberg in which unethical behavior is much more widespread. Many researchers are employed by companies or government bodies, and they are expected to come up with results that are useful to these funding bodies. As a simple example, those carrying out research on pesticides for chemical companies work within a restricted framework that does not include drawing attention to their dangers or limitations.

Martin (1992) has drawn attention to the power structure that exists in science and has detailed some cases where this leads to unethical behavior. The examples are restricted to the science scene in Australia, but it seems reasonable to assume that this is typical of what happens throughout the world. In exposing cases of unethical behavior, it should be noted that detailed descriptions are limited because of defamation law. Martin avoids examples that he refers to as misrepresentation of research progress, exploitation of subordinates, and bias in appointments, as it is considered that nothing will be done about them. He focuses on cases in which some action might be expected.

Cases

1. A person was appointed to a lectureship at an Australian university in preference to well-qualified applicants, based on an application that claimed that a PhD had nearly been completed at a prestigious overseas university. The PhD was never completed and it was later revealed that only a limited amount of work had been done. The appointee was supported by colleagues and was given tenure.
2. A lecturer who confronted his professor with evidence of the professor's plagiarism was physically threatened by the professor. After notifying the university administration of the evidence and action, the lecturer was transferred to another department against his will and nothing was done about the allegations.
3. William McBride is an Australian scientist who is famous for having discovered the link between the morning sickness drug

thalidomide and deformities in babies. Norman Swan of the Australian Broadcasting Commission reported allegations that McBride had falsified data in a paper that he had published in the *Australian Journal of Biological Sciences*. Phillip Vardy and Jill French, two junior researchers under McBride, had tried to raise their concerns with the directors of Foundation 41, where the research was done but got nowhere and resigned. In addition, seven other junior researchers contacted Foundation 41's Research Advisory Committee in regard to the allegations and they were subsequently retrenched. The case was made public as a result of the persistence of journalists Norman Swan and Bill Nicol. Foundation 41 set up an inquiry following the public disclosure and found that McBride had engaged in scientific fraud. Some time later, McBride returned to the Board of the Foundation.

Conclusions from these cases

Although only three of the cases presented by Martin have been mentioned in this chapter, these may be sufficient to illustrate some of the common threads that can be recognized where unethical conduct is exposed in science. First, it is apparent that unethical behavior occurs more abundantly than is usually assumed. However, its exposure to public scrutiny is relatively rare. One reason for this is that institutions are reluctant to take any formal action as this reflects negatively on their reputations. It is only when media attention is focused on a problem that they may take action. In many cases, this action may be to protect their interests rather than to find a solution to the ethical problem. The person who brings up the complaint frequently becomes a target, whereas the accused may find support. When the perpetrator resigns, the case is usually dropped. Very often, it is the whistle-blowers who finish up being disadvantaged most. They may lose their jobs or be moved to less attractive positions.

Fraud in science may be more prevalent than thought

Misconduct in medical research reveals numerous cases of fraud. Smith (2006) has described a number of cases and suggests that it is more widespread than commonly believed but is not easily dealt with. In this article, he discusses two Indian researchers who have published dozens of papers in major journals and expresses doubt about the authenticity of all of them. Hardly any of the studies have been retracted, as there is no one willing to take responsibility for investigating them. Rather ironically, the cases have been highlighted by the newspaper *The Wall Street Journal* and Canadian TV programs, but there has been little interest from the scientific community.

Misallocation of credit

Lawrence (2002) has described how the misallocation of credit for breakthroughs is endemic in science. Examples are given of how senior members of research groups are given credit for discoveries that have really been made by junior researchers. These junior researchers may slave away in the laboratory while their superiors travel and attend international conferences where they present the work of the group. In this way, it can be that important discoveries are attributed to them, whereas they are often made by the junior researcher. If the published work has several authors' names, it will usually be the senior one whose name is remembered. Lawrence cites several cases where this has occurred, in one instance resulting in the award of a Nobel Prize.

Maintaining the integrity of science

In order for science to progress, it requires researchers to be genuinely motivated by curiosity and a passionate desire to advance the frontiers of knowledge. They need to be prepared to overcome the obstacles and frustrations that they will inevitably encounter. It is not easy to discover new knowledge but, when this is achieved, the rewards can be great. We have briefly looked at examples of how scientists can stray from the high standards of integrity that they are expected to follow. There are obvious transgressions such as falsification or manipulation of data, plagiarism, and failure to acknowledge the work of others.

The emphasis on the number of publications as a criterion for performance evaluation tends to encourage researchers to put more effort into preparing publications to the detriment of perusing the scientific literature and of discovering significant new knowledge. The result has been an explosion in numbers of published papers as well as the numbers of journals. The consequence is that the quality of published work seems to be declining as pointed out by Bauerlein et al. (2010). It is vital that the integrity of science should be upheld and be supported by the public and this is more likely to happen if there is a sound practice of critical thinking and understanding of the scientific method. The problems that cause a lowering of quality need to be recognized and corrected. Misconduct needs to be openly addressed.

Authoritarianism

Another danger to scientific progress suggested by Popper is authoritarianism. Two simple examples of how authoritarian regimes have impacted science were mentioned in Chapter 2. These were the belief that the earth was the center of the universe, largely imposed by religion, and the dogma of Lysenkoism that was politically imposed. Authoritarianism is the opposite of scientific thinking. It is characterized by dogma, whereas science rejects dogma, always questions firmly held beliefs, and explores novel approaches to answering questions.

The earth as the center of the universe

For a long period, it was unquestioned that the earth was the center of the universe. The belief was reinforced by religious dogma. Eventually, it was overturned by scientific observations. We need to acknowledge, however, that science does not make religion untenable. For example, it is stated in the Bible that the world was created in six days, some few thousand years ago. In contrast, geological and chemical evidence shows that the earth, our most observable planet, has existed for several billion years. It has been argued that this disagrees with the biblical account and therefore the latter must be wrong. This is a rather naïve argument. Much of what is written in the Bible is symbolic and not intended to be taken literally. The failure to recognize this by both the religious and scientific communities has resulted in a lot of unfortunate misunderstanding.

Although doubts had been previously raised about the centrality of the earth in the universe, the studies of the Italian scientist Galileo Galilei were particularly effective in debunking this belief. Galileo used the newly invented telescope to make observations of the heavens. The early telescopes, which Galileo himself improved, made it possible to magnify distant objects by a factor of 20. The earth's moon was seen to be pitted with craters and the four main moons of Jupiter (now referred to as the Galilean moons) became visible. The planet Venus was seen to go through phases similar to the earth. All these observations were interpreted as showing the sun to hold a central position and the planets to revolve around it. This was consistent with the view that had been put forward earlier by Nicolaus Copernicus (1473–1543), but it added greater plausibility as a result of the experimental evidence. Galileo recorded his observations in a book titled *Siderius Nuncias* (the *Starry Messenger*). Subsequently, he was made to tread a fine line in regard to the teachings of the church. He was permitted to continue his writings on the condition that he treat the Copernican theory as only hypothetical. Unfortunately, it was later determined that he had breached this condition and, as a result, a case was brought against him by the Inquisition. He was pronounced to be suspect of heresy and made to admit that he had overstated his case, and was forced to formally recant.

Lysenkoism

Beginning in the late 1920s, Trofim Lysenko, who became a director in the Soviet Union's Academy of Agriculture, introduced a theory built on the premise that characteristics acquired by environmental factors are heritable. This went against theories that were being developed in other parts of the world based on the inheritance of traits that were determined by genetics, following the work of Gregor Mendel.

Lysenko aligned himself with the political orientation of the Soviet Union and entered the hierarchy of its Communist Party. The backing of the party and its leaders enabled him to exert a considerable influence on science, particularly in the area of agriculture. He was able to hide the falsity of his theories by devious methods, including appearing to solve one problem and then quickly moving on to others. He became a hero and the powerful propaganda machine of the government protected him from close scrutiny. Many of his colleagues profited by supporting his ideas. Those who voiced their disagreement were often dismissed from their positions and, in some cases, imprisoned. This is the state of affairs that invariably occurs when an authoritarian regime is imposed and this is particularly the case in science. Those who are opportunistic, flourish while those with integrity are disadvantaged.

Managerialism

Some of the negative effects of excessive management control on science have already been discussed in Chapters 3 and 4. It has only been in the past few decades that managerialism has been seriously imposed on science. Just as an authoritarian political regime curtails the freedoms of citizens, the imposition of a managerial system restricts scientists in utilizing their creative talents. In Chapter 3, we compared two procedures for carrying out scientific research, which were termed the scientific and managerial approaches. The managerial approach sets objectives and a well-delineated schedule for achieving them. The scientific approach begins with problems in mind but, rather than pursuing a predetermined plan to achieve strict goals, allows ideas to evolve and to pursue promising ones wherever they may lead. Of course, this is not a black-and-white issue. A certain amount of control is needed to keep research focused on the problems that are to be addressed. It is when this control becomes excessive that scientific inquiry is stifled.

One important difference between the two approaches may be summed up by the word *imagination*. In the managerial approach, there is little room for imagination. The objectives and the paths to achieve them are mapped out from the beginning. In contrast, the scientific approach allows development of imaginative ideas that open up possibilities for unexpected discoveries. These discoveries are often ones that lead to spectacular breakthroughs. It is worth remembering the words of Albert Einstein that have been previously mentioned: "Imagination is more important than knowledge. For knowledge is limited to all we know and understand while imagination embraces the whole world and all there ever will be to know and understand." This statement captures the essential difference between the two approaches we have been considering. The managerial approach is destined to discover nothing more than what had been planned at the outset. The scientific approach, by contrast, is not limited to only the initial expectations but, through imaginative thinking, can soar into realms of understanding that had previously not been dreamed of.

References

Alberts, B., and Shine, K. 1994. Scientists and the integrity of research. *Science* 266:1660–1661.

Bauerlein, M., Gad-el-Hak, M., Grody, W., McKelvey, B., and Trimble, S.W. 2010. We must stop this avalanche of low-quality research. *The Chronicle of Higher Education*, June 11.

Behbehani, A.M. 1983. The smallpox story: Life and death of an old disease. *Microbiological Reviews* 47(4):455–509.

Boon, S. 2016. 21st century science overload. *The CSP Blog*, January 7. http://www.cdnsciencepub.com/blog/21st-century-science-overload.aspx.

Evans, J.A. 2008. Electronic publication and the narrowing of science and scholarship. *Science* 321:395–399.

Hamilton, D.P. 1990. Publishing by—and for?—the numbers. *Science* 250:1321–1332.

Holden, C. 1991. Do we need more Ph.D.s, or is fewer better? *Science* 251:1017–1018.

Lariviere, V., Gingrus, Y., and Archamboult, E. 2009. The decline in the concentration of citations, 1900–2007. *Journal of the Association for Information Science and Technology*, January 29.

Lawrence, P.A. 2002. Rank injustice: The misallocation of credit is endemic in science. *Nature* 415(6874):835–836.

Martin, B. 1992. Scientific fraud and the power structure of science. *Promethius* 10:83–98.

Smith, R. 2006. Research misconduct: The poisoning of the well. *Journal of the Royal Society of Medicine* 99:232–237.

Squires, B.P. 1992. Citation rate: A measure of excellence? *Canadian Medical Association Journal* 146(3):341.

chapter seven

Applying scientific thinking to some current controversies

We are going to look at several issues that have been the subject of controversy. These are issues that require a scientific approach, although those who participate in the ensuing debate certainly need not necessarily be restricted to scientists. In some of these cases that we are going to examine, large sections of the public have been persuaded to accept a certain viewpoint on the basis that it is supported by science. For each one, we will apply criteria described in earlier chapters in order to assess whether the suggested viewpoint or theory is justifiably based on scientific thinking. It is not the intention to agree or disagree with the theories but only to critically examine the authenticity of the scientific analysis.

Climate change

As pointed out early in the book (Chapter 1), the title of climate change is totally ambiguous in relation to the current controversy. There should be no doubt that the earth's climate has changed in the past, is changing currently, and will likely change in the future. The issue that has stimulated debate is that of anthropogenic global warming, that is, that the emission of greenhouse gases (of which carbon dioxide is one) from the planet's surface is making a significant contribution to its warming. These gases emanate from various sources, but the main one that is proposed is from industrial processes such as burning of fossil fuels, mainly coal and petroleum. Another source of ambiguity in the debate stems from the mistaken adoption of other terms such as "carbon" and "carbon pollution" to describe the emission of greenhouse gases. Carbon dioxide is an essential compound for plant growth and certainly should not be classed as a pollutant.

First, we should dismiss the term "climate change" and accept the term "anthropogenic global warming" as the topic of debate. We frequently hear expressions such as "climate change is real." Everyone agrees with that. Then, it is widely stated that the science of global warming is "settled," implying that it has been "proven" and therefore should not be questioned. It seems remarkable that such an obvious misconception of the scientific process could be so widely propagated, even by those who have scientific qualifications and call themselves scientists. One of the fundamental tenets of science is that no theory is ever proven beyond doubt, as has been explained previously, particularly in Chapter 2.

Observations consistent with previous ones cannot be claimed as confirmations

The claims for accepting the theory of anthropogenic global warming as proven are based on observations that appear to confirm it. For example, sea levels have been reported to be rising in some parts, an effect that would be expected if the earth (or oceans) was warming. Other confirmatory observations are that the amount of ice in the arctic region is decreasing and that glaciers are melting in some parts of the world. Each of these observations is claimed by the proponents of the theory as additional confirmation of the theory. This is a characteristic of the inductive method, which has been shown to not be a valid procedure for scientific discovery. Induction is an example of the dogmatic attitude, which is to look for confirmations of a theory. Confirmations are easy to find if we look for confirmations. The true method for discovering scientific knowledge is that of hypothesis-deduction. Instead of waiting passively for observations of nature to be made and then attempting to arrive at a theory as in the inductive method, the hypothetico-deductive method makes inspired guesses or conjectures and then designs experiments to severely test them with the intention to obtain a possible refutation. There is therefore a fundamental difference between the two methods. Whereas induction proceeds from observation to theory, the hypothetico-deductive method proceeds from theory to observation. The aim with induction is to try to confirm a theory. In contrast, the aim in hypothesis-deduction is to propose a theory and then subject it to severe tests that could result in a possible refutation. If severe tests of the theory fail to refute it, the theory is corroborated and tentatively held to be true, although it can never be considered to be proven. In the evolution of science, the transition from a dogmatic approach (induction) to a critical approach (hypothesis-deduction) represents a major advance in scientific thinking.

The weakness of modeling

There do not seem to have been any experiments designed to try to refute the theory of anthropogenic global warming so the hypothetico-deductive procedure has not been seriously applied. Perhaps the approach that has come closest is that of modeling. Modeling does, at least, attempt to predict future temperature events. However, a good hypothesis is one that is based on a new and simple concept, and this is not what is involved in the modeling that has been applied to the question. The arguments in favor of global warming have focused on confirmations, the inductive, and thus the dogmatic approach. It will be interesting to see if its proponents can come up with experiments that could, in principle, refute the theory. If not, its supporters may claim that it is irrefutable. However, as we have seen, irrefutability is not a virtue of a theory but renders the theory nonscientific. Even if it can be demonstrated that there is warming of the planet on a global scale, there still remains the question of whether it is due to human activity or to natural climate variability. This question has been mainly addressed by applying modeling. In this activity, an input is created in which all the variables that are expected to influence climate are fed in. The output from the model then gives an answer to how the climate will be affected. Modeling works well providing the problem being tackled is fairly simple and all the variables are understood and properly introduced into the model. In the case of climate change, this is an extremely complex problem. It is doubtful if all the variables are adequately recognized. If they are not, the exercise becomes one best described by garbage in, garbage out.

There is another weakness in the use of modeling to predict climate change. We can observe how the climate is changing so there could be a temptation to manipulate the inputs to satisfy the outputs. There is a big difference to the case of the prediction from Einstein's theory that light should be deflected by large bodies (see Chapter 2). This was a risky prediction of an event that had not been observed. As mentioned before, many participants in the debate would not contemplate deliberately attempting to disprove their dogmatically held opinion.

The medieval warm period

An observation that could possibly be considered as a test of the theory relates to what has been referred to as the Medieval Warm Period (MWP). This was a period between the ninth and fifteenth centuries in which the rate of warming and the temperatures attained in parts of the Northern Hemisphere are postulated to be similar to what has been observed globally in recent decades. A proposal has been suggested that a study be made of fossils, cave deposits, and tree ring records from tropical to Antarctic Australia to test whether the warming evidence could be shown to be global (Asten, 2010). If this were shown to be the case, then the belief that human contributions are causing today's warming would be undermined since it would show that warming periods similar to what is experienced today occurred when carbon dioxide was not being emitted by industrial processes. If it were shown to not be the case, it could be taken as a failed refutation and the theory of human effects would be corroborated and emerge stronger. Studies in the Southern Hemisphere have in fact been reported suggesting that the Medieval Warm Period was global.

A study by an international research team (Neukom et al., 2011) reconstructed a mean temperature history for the period 900 to 1995 for a region they called Southern South America (SSA). It was found that a warm period extended in SSA from 900 AD or earlier to the mid-fourteenth century. The warmest period of the MWP was calculated to be from 1079 AD to 1089 AD and the temperature was estimated to be about 0.17°C warmer than the peak warmth of the current warm period. Another study (Bracco et al., 2011), based on phytoliths (microscopic plant fossils), similarly concluded that Uruguay's climate of 750 AD to 1350 AD was warmer than the current climate. The intention here is to show that it is this type of approach (hypothesis-deduction) that is needed to advance knowledge on the subject and not the dogmatic approach based on induction.

One of the arguments against ascribing warming to emissions of greenhouse gases is that there has not been any measurable increase in atmospheric temperatures for a recent period of about eighteen years or so. This is despite the fact that industrial emissions have been continually increasing. Confronted with this apparent anomaly, the proponents of anthropogenic global warming are searching for explanations. One that has been suggested is that the extra heat produced by industry is absorbed into the ocean depths. When a hypothesis is refuted, there are two possible courses of action. The first one is to accept the refutation, to admit that the hypothesis was wrong, and to proceed to form a new hypothesis. The risk in adopting this conclusion is that it may be premature and that, if the theory were to be persevered with, new evidence or new interpretations of the evidence may arise that supports the theory. The second course of action is to introduce auxiliary hypotheses to modify the original hypothesis so as to try to make it fit the observations. This can certainly always be done, but the introduction of auxiliary hypotheses usually signifies a lowering in scientific status of the theory.

The Great Barrier Reef

The Great Barrier Reef is a coral reef system that stretches for over 2300 kilometers (1400 miles), located in the Coral Sea off the east coast of the Australian state of Queensland. It has been built up from billions of tiny organisms, the coral polyps. It is the world's largest structure made by living organisms and can be seen from outer space. It attracts large numbers of tourists, and the protection of a large part of it is administered by the Great Barrier Reef Marine Park Authority. This area is home to many species of fish and other marine animals. Since 1981, it has been classed as a World Heritage Site. As a result of the delicate nature of its structure, the reef is acutely susceptible to environmental effects. Its uniqueness and touristic attraction has thus been a source of concern for its preservation. The threats to the health of the reef come from both human and natural causes. It has been suggested that the human causes arise from the presence of a number of ports along the coast with their associated shipping and from mining operations in the vicinity.

Similarly to the global warming debate, there has tended to be a polarization of opinion in regard to the extent of damage that is occurring and its causes. On the one hand, those who are predominantly concerned from the environmental side have emphasized the dangers to survival of the reef, whereas on the other hand the industrialists have argued against the effects of their activities. Here, we will only give a brief survey of the debate to focus on the science behind some of the arguments that have been presented, without favoring any overall conclusion.

There are a number of effects that are capable of placing the reef in danger. As mentioned, some of these are natural while others stem from human activities. The natural ones include the crown-of-thorns starfish, an organism that feeds on coral polyps. Another is coral bleaching, a cyclical event that causes loss in color of the coral. Effects resulting from human activities include pollution of the water due to farm runoff, which may deliver chemicals such as pesticides and heavy metals such as lead, mercury, and arsenic. These effects can be aggravated by overgrazing in agricultural land adjacent to the coast and intensive cropping such as may result from growing of sugar cane. Other pollutants could be dispersed into the waters from mining operations near the coast.

Coral bleaching effects are known to be promoted by increasing temperature. Thus, the theory of anthropogenic global warming has been linked to bleaching in the reef. Bleaching events have been observed in coral systems around the planet. In recent times, they occur rapidly and it takes long periods for regeneration to occur. As a result, there have been warnings that coral ecosystems are likely to disappear in the relatively near future. These warnings have been exacerbated by pronouncements of politicians such as were made recently by a United States president in an address to students at a Queensland university. It is unfortunate that politicians who wield a great amount of influence on public thinking enter into debates that are outside their areas of expertise.

Effects of acidification of the ocean

As mentioned, there has been a halt in warming of the earth's atmosphere in a recent period of 18 years or so. This has caused environmentalists to search for alternative mechanisms to relate degenerative processes in the reef, such as bleaching, to human activities. One suggestion is that the increase of carbon dioxide in the atmosphere should lead to its greater solubilization in the ocean, causing acidification of the water. This arises because dissolution of carbon dioxide produces formation of the weak acid, carbonic acid, with a lowering of pH or, in other words, an increase of acidity. It has been suggested that increasing acidity might dissolve the coral reefs and kill organisms with calcareous shells such as oysters, clams, and mussels.

The chemistry of carbon dioxide dissolution

The suggestion that increased carbon dioxide and the resulting atmospheric warming will lead to acidification of the oceans is rather simplistic and needs to be considered in relation to the basic chemistry involved. Carbon dioxide is soluble in water but, unlike some other gases such as oxygen, it reacts to form a balance of several chemical species—dissolved free carbon dioxide, carbonic acid, bicarbonate ions, and carbonate ions. The chemical balance of these species needs to be considered when trying to explain processes such as calcification. One simple result that needs to be realized is that, although emissions of carbon dioxide are expected to cause atmospheric warming, the solubility of carbon dioxide decreases with increasing temperature. An increase of atmospheric carbon dioxide would therefore tend to slow responses to emissions.

Another factor to consider is that the high concentrations of basic elements such as calcium and magnesium have a buffering effect on the seawater, that is, they stabilize the pH, thus preventing large fluctuations in acidity.

A critical examination is needed of the arguments for and against

One outcome of the debate on the sustainability of the Great Barrier Reef is the polarization it has caused among those who participate in the debate. As in the case of the debate on global warming, there has been a sharp separation between those who assert that human activities are causing serious damage to the reef and those who believe that this view is greatly exaggerated. Industrialists tend to smooth over the influences of their activities. Environmentalists, on the other hand, tend to exaggerate effects that tie into their agendas. This can lead to outright falsifications. As just one example, a composite image of two photographs taken just six months apart purportedly to show dramatic coral bleaching was found to have been taken thousands of kilometers from the reef (Passmore, 2016).

The adoption of dogmatic stances on issues is opposed to what should happen in a debate that requires application of scientific thinking. The correct position that should be taken on an issue that is complex and in which there is conflicting evidence is to withhold one's judgment until more solid evidence is acquired. People need to be acutely aware of the bias that is likely to be present in the media and not to believe everything they hear or read.

The saturated fat controversy

One of the current controversies in the difficult area of medical research is that of a link between saturated fats in the diet and heart disease. Fats consist of molecules with long hydrocarbon chains, that is, chains containing hydrogen and carbon atoms. In saturated fats, molecules have only single bonds between hydrogen and carbon so that there is a maximum ratio of hydrogen to carbon. Unsaturated fats have lower ratios of hydrogen to carbon because some of the bonds between the two elements are double bonds. For several decades, the view has been generally held that saturated fat is a risk factor for cardiovascular disease (CVD). This view has been pushed in the health guidelines of many institutions, including the American Heart Foundation, the British Heart Foundation, the National Heart Foundation of Australia, the Heart Foundation of New Zealand, and the Heart and Stroke Foundation of Canada. In recent times, the view has been challenged and its status is the subject of robust debate.

The topic has generated a huge amount of research. It is not practical to give an exhaustive review of this research here. What will be done is to present a brief summary of the main conclusions and to point out the complexity of the problem and some of the inherent difficulties in this type of research. The general aim in all research is to study the effect of one variable while attempting to eliminate or, at least, to identify all other variables that might have an impact. In the case of measurements involving human subjects, this is not easy. All human beings are different from each other. This includes their genetic makeup, their lifestyle, and their responses to different environmental effects as well as their capacity for close cooperation in trials. If, for example, we are measuring the effects of a specific diet on a particular aspect of health, all of these variables play a role. The experiments are designed to try to minimize these variables but a certain level of uncertainty always remains.

Brief summary of the research on the effects of saturated fats

Ancel Keys introduced the hypothesis that fats in the diet, in particular saturated fats (as occur in meat and dairy products), increased the risk of heart disease. He began his studies at the University of Minnesota. Later, he traveled widely and enlisted the help of scientists in different countries. This led to the seven-country study in which surveys appeared to show a relation between consumption of fats and heart disease (Keys, 1980). This result was publicized in the 1950s. It was found that saturated fat consumption increased total cholesterol levels in the blood and this was adopted as an important marker for heart disease. Other factors such as high blood pressure and smoking were identified as risk factors, but we will be concerned here only with diet. The powerful bodies of the American Heart Association (AHA) and the National Institutes of Health (NIH) adopted the diet–heart hypothesis of Keys and introduced recommendations to reduce the amount of fats (and saturated fats) and replace them with polyunsaturated oils such as corn or soybean oil. The Department of Agriculture (USDA) introduced dietary guidelines in the form of a pyramid. This pyramid emphasized the healthy nature of foods high in carbohydrates (breads, cereals, vegetables, and fruit), while fats and oils were relegated to the top smaller section of the pyramid.

Problems for food processors

The recommendations to decrease saturated fats in the diet have been in place for about half a century in the United States and other parts of the world. The push to decrease the consumption of saturated fats has created a problem for food processing, particularly for the large manufacturers. Saturated fats such as lard and tallow are solid and impart firm texture and good keeping properties to products. Polyunsaturated oils are liquid and do not provide the textural properties required for many processed goods. They are also less stable and can give rise to oxidation products that are not desirable. Food scientists, in order to deal with this challenge, have adopted the process of partial hydrogenation in which the ratio of hydrogen to carbon is increased in the hydrocarbon chains. This is a chemical process in which the properties of the unsaturated oils are shifted toward those of saturated fats. This gives them more solid texture and improves the keeping properties. An example is margarine in which an unsaturated oil is converted to a more solid product.

The problem of trans fats

Although partial hydrogenation has solved one problem for food processors in that it provides fats with suitable texture and stability, a new problem has emerged. The hydrogenation process (involving the introduction of hydrogen atoms into the hydrocarbon chains) occurs at points of unsaturation (called double bonds). This gives rise to a change in configuration of the chain from cis to trans. The presence of trans fats has been related to increase in heart disease. A prominent scientist whose work has shown this is W.C. Willett (Willett and Ascherio, 1994). As a result, trans fats are being phased out and food processors are being obliged to seek alternatives.

Replacement of trans fats

One of the alternatives to using trans fats, and because of the bias that has built up against saturated fats, is simply to use polyunsaturated vegetable oils such as those from soybean or corn. The problem with these oils, apart from their texture, is that they are unstable. When they are heated (as in frying), they may break down to give toxic products such as aldehydes, which present health problems. Olive oil is a monounsaturated oil (i.e., has only one double bond in its fatty acid chain) and is therefore more stable than the polyunsaturates. It is a component of the Mediterranean diet that has been associated with good health. Another avenue being pursued is to develop fat replacers. These aim to substitute for the textural properties of fats and are essentially made up of carbohydrates.

The role of cholesterol

In the early days of the saturated fat hypothesis proposed by Keys, total cholesterol level was found to be raised by saturated fat and so was thought to be a risk factor in heart disease. Medical advice recommended that it should be monitored and lowered. This could be done by eliminating foods high in cholesterol such as eggs from the diet or by special drugs such as statins. The truth is that cholesterol, far from being an unhealthy compound, plays an essential role in the body. Its molecule consists of three parts: a hydrophobic (fat-soluble) hydrocarbon chain at one end, a hydrophilic (water-soluble) hydroxyl group at the other end, and a set of four hydrocarbon rings between the two ends.

This structure fits it to play an important role in building membranes and modulating their fluidity, thus controlling the flow of nutrients into the tissues. Cholesterol is not soluble in the blood and, together with fats such as triglycerides, needs to be incorporated in lipoprotein bodies to enable it to be transported via the blood to the cells.

LDL and HDL cholesterol

The lipoprotein bodies that carry cholesterol and fats have been classified as low-density (LDL) and high-density (HDL) lipoproteins. An oversimplified characterization has been to refer to LDL cholesterol as "bad cholesterol" and HDL cholesterol as "good cholesterol." Recent research has shown that these two fractions are more reliable markers for heart disease and that total cholesterol is not a useful marker. It is believed that the LDL cholesterol tends to deposit on artery walls to form plaques that constrict blood flow, whereas the HDL cholesterol scavenges the walls, removing excess cholesterol and returning it to the liver.

A critical examination of the saturated fat–heart disease hypothesis

What do we find when we apply critical scientific analysis to evaluate the huge amount of work that has been directed to the cause of heart disease, an ailment that has become one of the major life-threatening diseases afflicting Western countries in the last century? It becomes apparent that there have been some flaws in the way science has been applied.

Epidemiological research versus clinical studies

The first thing to realize is that the hypothesis of Keys was almost exclusively based on epidemiological research. This type of observational research studies the diet of groups (cohorts) of similar individuals (e.g., males of a certain age) and relates the diet to health effects such as heart disease. The weakness of this approach is that, although correlations may be found, these do not necessarily signify cause-effect relationships, as was always admitted by Keys. Cause-effect relationships can only be determined using clinical trials in which an intervention is introduced and the effects on health are measured. The intervention may be a change in diet or it may be the administration of a certain drug. All the research on which the initial saturated fat–heart disease hypothesis was proposed was based on epidemiological studies.

Selection bias

Once the saturated fat–heart disease hypothesis was formulated and strongly promulgated by its proponents, subsequent research tended to be self-fulfilling. Studies that appeared to refute the fat–heart hypothesis were played down. Those showing that certain groups of people consuming diets high in saturated fats who did not suffer from heart disease (e.g., the Masai people of Kenya) were ignored. These are examples where selection bias operates. This has become a common scientific failing where those who strongly support a certain theory selectively ignore contrary evidence and where experiments not supporting their theory are not emphasized. These deviations from honest application of the scientific method in the case of the fat–heart hypothesis have been illustrated in the well-researched book *The Big Fat Surprise* by Nina Teicholz (2014). This author has no allegiance to any vested interest in the issue and this enabled her to provide a fair and objective history of research on the topic.

Encroachment of bureaucracy

Once the powerful American institutions such as the AMA (American Medical Association), NIH, and USDA, aided by the media, became involved in the issue and supported the fat–heart hypothesis even though it had not been adequately tested, scientific criticism became stifled. As Teicholz writes (p.103) in relation to criticisms by experts:

> Whatever chance these experts might have had for self-correction was lost, however, when the federal government got involved. With its massive bureaucracies and obedient chains of command, Washington is the very opposite of the kind of place where scepticism—an essential to good science—can survive. When Congress adopted the diet–heart hypothesis, the idea gained ascendancy as an all-ruling, unassailable dogma, and from this point on, there has been virtually no turning back.

Status of the hypothesis

It has now been a half-century since people have been advised to cut down on fat, and especially saturated fat, in their diet. Therefore, we should be in a position to make some observations about the effect it has had on health. In the interim, there have been important medical advances in heart care, apart from diet, that need to be factored in. Heart disease remains high on the list of serious ailments. Other observations are that there has been a spectacular increase in obesity and diabetes in some countries including the United States. The question being asked is whether this has been caused by adoption of a low fat diet. There are three main macronutrients in the human diet: proteins, carbohydrates, and fats. Reduction of fats in the diet means that they must be replaced by other nutrients. This has been mainly carbohydrates and this is what is being suspected of contributing to the increase in obesity and diabetes.

Some of the doubts about the validity of the hypothesis have been noted. Teicholz (2014), in her excellent book, has pointed out many examples where saturated fats in the diet have either not been shown to be detrimental to heart health or have been beneficial. It is notable that proponents of the saturated fat–heart disease hypothesis have adopted a rather curious position. That is that, having put the hypothesis in place without adequately testing it, they appear to assert that it is up to its opponents to prove it wrong. The results of the roughly half century of testing the hypothesis have cast considerable doubt on its validity.

The advent of gluten-free foods

We have seen (in the previous section) how the public has come to accept a belief that fat in the diet, and especially saturated fat, is detrimental to heart health. This has not occurred because the theory is based on sound science. The theory has come to be accepted to a large extent through selection bias and by powerful bureaucracies that have promoted it. A similar thing is happening with the current explosion in gluten-free foods, based on the concept that wheat and its component gluten are bad for health. There is, nevertheless, some firm basis for this theory.

Celiac disease and gluten intolerance

A small percentage of the population (roughly 1–2 percent, but this varies in different countries) suffers from a serious autoimmune disorder called celiac disease. Autoimmune disorders are ones in which the body perceives certain ingested nutrients as alien invaders and marshals its resources in the form of autoantibodies to produce an inflammatory reaction to attack its own cells. Celiac disease is caused by eating gluten, a protein that occurs in several cereals such as wheat, rye, and barley. It causes damage to the lining of the small intestine and leads to serious health problems resulting from preventing the absorption of essential nutrients. It is a debilitating disease that can be most reliably diagnosed by carrying out an intestinal biopsy. If it is neglected, it can have long-term consequences such as anemia, osteoporosis, and bowel cancer. The only known way at present for celiac patients to be free of the disease is to eliminate all gluten from the diet. This is not so easy to do. There are foods that obviously contain gluten, such as bread and pasta from wheat. However, there are many other foods such as sauces that contain sufficient amounts of gluten to make them unacceptable for celiac patients.

There is also a small percentage of the population who, although not celiac patients, has an intolerance to wheat or to gluten. Ingestion of wheat or gluten can produce ailments such as allergic reactions.

The need for gluten-free foods

Although the percentage of people affected by celiac disease and gluten intolerance is small, its numbers are significant. For example, in the United States, it is estimated that there are about 1.8 million who suffer from celiac disease. This has stimulated research into developing foods for this niche market. There are cereals that do not contain gluten (e.g., maize, rice, and sorghum). These have been utilized to try to find substitutes for the traditional gluten products. A fundamental problem is that wheat and rye (but to a lesser extent) are the only cereals that form dough having viscoelastic properties. It is these properties that allow dough to stretch and encapsulate tiny air bubbles. Baking of the aerated dough then produces the light, palatable products such as breads that are highly prized by consumers.

Explosion of gluten-free foods

In addition to the estimated 1.8 million people in the United States who suffer from celiac disease, there is also a small percentage of the population that has what goes under the name of gluten sensitivity. Unlike celiac disease, which can be accurately diagnosed, gluten sensitivity is difficult to measure. As a result, there are people who genuinely improve their health by avoiding gluten, but there appears to be many for whom the elimination of gluten is simply a fad. It has been reported that 1.6 million people in the United States are on a gluten-free diet even though they have not been diagnosed with a gluten problem. A third of adults in the United States have expressed a wish to cut down or eliminate gluten consumption. We are left to conclude that only a small percentage of the population incur negative effects to their health by consuming gluten. In contrast, there has been an explosion in the quantities of gluten-free foods available on the market.

It has been estimated that the market in gluten-free foods is expected to increase in value from $1.27 billion in 2013 to close to $24 billion by 2020. Apart from catering for the niche market of celiac and gluten-sensitive patients, what is the basis for the projected increase in gluten-free foods? Just as was the case for the saturated fat–heart disease theory, it does not seem to have a solid scientific basis. The decision to adopt gluten-free products is influenced by the strong marketing programs of manufacturers and by the pronouncements of certain celebrities.

Many people find that they feel better when they make changes to their lifestyle and diet. They may attribute this to reducing or eliminating gluten, but, in many cases, the basis for the belief may not be valid. In order to establish a cause-effect relationship, controlled experiments are required in which only one variable at a time is changed. Often, when people make changes to their lifestyle they may change many variables such as exercising more or other changes in diet including a reduction in total calorie intake, so it is risky to identify one of these variables as the cause of their feeling better.

The downside of gluten-free diets

Wheat has been the most widely grown cereal crop. It is the staple food for some 35 percent of the world population and possibly accounts for about 70 percent of the total calorie intake for some of the developing countries.

Thus, the negative effects that are being attributed to wheat and its gluten, should they continue to be exaggerated, pose a threat to a large industry and, ultimately, to world food security. The trend toward reduced acceptability of wheat and gluten in the general population (i.e., that is free from celiac disease and gluten sensitivity) has not been driven by science-based evidence. It has not been so much the result of advice from dietitians and medical practitioners. It has been promulgated more by articles in the popular press and by advertising campaigns directed at selling gluten-free products.

Furthermore, it has been found that gluten-free diets can be seriously deficient in certain nutrients. Many gluten-free foods are made from refined and unenriched grains and starches. These may be high in calories but low in important nutrients such as minerals, vitamins, and fiber (Shewry, 2016). Thus, there is a potential risk to health.

Genetic engineering

The past few decades have seen a controversy about the introduction of genetic engineering (GE) to the breeding of plants and animals. Traditional or conventional breeding has involved mating of animals and sexual crossing of plants, whereas GE uses artificial manipulation. The aim in traditional breeding is to rearrange the many genes provided by the parents in order to accentuate desirable traits. Each trait or characteristic of a plant or animal is determined by a gene. For example, in the case of plants, the desirable trait might be an increased resistance to pests or a greater tolerance to drought.

The hereditary component of genes is DNA (deoxyribonucleic acid), mainly present in the nuclei of cells. DNA is a large polymeric molecule made up of four different units or bases strung together in different sequences. It is the order in which the bases are arranged that determines a specific trait in a similar way to how letters of the alphabet appear in a certain order to form words and sentences. As a result of work mainly by Franklin, Watson, Crick, and Wilkins (Watson and Crick, 1953), the DNA molecule was determined to exist as two coiled strands in the form of a double helix. Each strand has the capacity to replicate. Before a cell divides, the strands separate and each forms complementary strands for incorporation into new molecules, thus preserving the genetic code.

The possibilities for genetic manipulation

Subsequent studies have shown how DNA could be manipulated. For example, it could be split at certain points and sequences of other DNA could be inserted, thus altering the genetic blueprint. Techniques were developed for isolating specific genes and introducing them into organisms or, alternatively, deleting an undesirable gene. These new techniques represented powerful tools for modifying organisms. They allowed, at least in theory, for traits to be introduced by insertion of single genes or DNA sequences, rather than having to rely on conventional breeding methods, in which many genes are rearranged, followed by a lengthy process of selection and backcrossing. The breeding process would thus be speeded up. Furthermore, the genetic material to be inserted was not restricted by species boundaries as in conventional breeding. Genes could be introduced into an organism from a different species.

Nowadays, there is a wider range of tools available. As well as sequencing and synthesizing, techniques are available to cut sections of DNA with increased precision. DNA from other species can be inserted and genes can be turned on and off. As an example of the advances, the insulin gene has been inserted into the *Escherichia coli* bacterium and insulin produced in vats of E. coli. This has allowed a clean and consistent production, avoiding the need to rely on animal tissue (pig pancreas). Similar procedures are being used to produce vaccines and medicines for targeting specific diseases.

Gene drives

A recent development is the concept of gene drives. A gene drive uses a procedure in which a gene and its associated trait are passed on to all subsequent generations (Nolan and Crisanti, 2017). As an example of its potential use, gene drives could be introduced into mosquitoes to make either the male or female sterile. This would be an effective way of reducing or eliminating the transmission of diseases such as malaria, which take a heavy toll on people in certain regions of the world.

CRISPR

Another technique that is being developed is CRISPR (clustered regularly interspaced short palindromic repeats). This technique can be used to edit genomes, a genome being a complete set of genes or genetic material present in a cell or organism. Thus, the cell's genome can be cut at a desired location, allowing existing genes to be removed and/or new genes to be inserted. CRISPR has the potential to revolutionize areas of medicine and crop seed enhancement.

Weighing the risks

The brief summary of genetic engineering (GE) that has been given provides a glimpse of the promise of this technology to provide benefits.

It has many potential applications in areas such as agriculture, food production, and medicine. As in all new technologies, however, there are risks that need to be considered. A debate has ensued in which many scientists argue that GE is simply an extension of the science that has used genetic manipulation in conventional breeding. The opposing argument is that it conflicts with nature and this inherent risk requires that it should not be adopted or, at least, that much caution be exercised prior to its application. The two points of view are illustrated by the different approaches that have been taken by the United States and Europe in the growing of genetically modified (GM) crops. Large proportions of crops such as corn, soybean, and cotton grown in the United States are genetically modified. In Europe, there has been a greater resistance to planting of GM crops and a greater insistence that foods made from them should be suitably labeled as such.

What are the risks?

Some of the risks that GE may pose can be easily envisioned. GM crops can move proteins into the food supply from sources that have never been consumed as foods. There is a risk that these proteins can be allergens.

There is always the danger that novel genes that are introduced into crops can be transferred via pollen to related plants outside the agricultural fields. A possible result is then that the new traits can cause these plants to thrive and become unwanted "superweeds." Similarly, traditional organic crops can be contaminated by GM crops in adjacent fields.

Herbicides are sprayed on crops to fend off insects but can be harmful to the plants. GM crops have been developed to have high tolerance to herbicides. Commonly used herbicides contain compounds that have been designated as probably carcinogenic, thus their excessive use could present a possible health danger for consumers. Another danger is that the herbicide-resistant trait could be passed from the GM crop to weedy relatives, resulting in their unwelcome proliferation.

Seed companies can take out patents on seeds that have been genetically modified. One concern is that these companies could restrict the use of GM crops by causing second-generation seeds to be sterile. These are referred to as "terminator seeds." There would then be a need to buy seeds for planting each season, an imposition on farmers and, in particular, smaller landholders. Fortunately, although this technology is available, it has so far not been applied commercially.

Techniques such as CRISPR, described earlier, make heritable changes to genomes and there are worries that these could disrupt entire ecosystems. These areas of research are advancing at such pace that there are concerns that insufficient time is being allowed to address the ethical and safety issues that they raise (Ledford, 2015).

A critical examination of the controversy

We have looked at a few of the potential benefits of GE and at a few of the potential risks. GE is a relatively new technology so that some of the benefits and the risks cannot be envisaged and may only become evident as the research advances. As in the other controversies we have considered, there are two sides of the argument and many have aligned with one side. This can lead to biased thinking.

There has been a tendency, particularly in the early days of GE research, to overstate the expected benefits. There have been suggestions that GE would solve the world's hunger problems. In recent times, there is recognition that this claim needs to be downplayed. Certainly, there are benefits to be had, but there is a need for them to be realistic. Many factors contribute to the lack of food availability for populations, including its transport, capacity of people to pay, and the wastage.

Many who oppose the application of GE have also shown biased thinking. Trial plots of GE crops have often been vandalized by activists. As an example, a field trial of "golden rice" was destroyed by uprooting of all the plants by a group of some 400 protesters in the Philippines (Kupferschmidt, 2013). Traditional rice varieties possess the mechanism for synthesizing beta-carotene, the precursor of vitamin A, in the leaves but not in the grain. By adding a daffodil gene and a bacterium gene, the pathway is turned on so that beta-carotene accumulates in the grain, giving it a golden color. Vitamin A deficiency is widespread, particularly in poorer countries, causing chronic health problems. The aim of the research in the Philippines was to develop a rice variety to alleviate the vitamin A deficiency. A project with noble objectives was therefore disrupted by people whose actions were based on ignorance and irrational prejudice.

References

Asten, M. 2010. CSIRO should establish if there was medieval warming down-under. *The Australian*, May 13. www.theaustralian.com.au/news/.../story-e6frg6zo-1225865724876.

Bracco, R., del Puerto, L., Inda, H., Paniero, D., Castineira, C., and Garcia-Rodriguez, F. 2011. The relationship between emergence of mound builders in SE Uruguay and climate change inferred from opal phytolith records. *Quaternary International* 245:62–73.

Keys, A. 1980. *Seven Countries: A Multivariate Analysis of Death and Coronary Disease.* Harvard University Press, Cambridge, MA.

Kupferschmidt, K. 2013. Activists destroy "golden rice" field trial. *Science,* August 9.

Ledford, H. 2015. CRISPR, the disruptor. *Nature,* June 3.

Neukom, R., Luterbacher, J., Villalba, R. *et al.* 2011. Multiproxy summer and winter surface air temperature field reconstruction for southern South America covering the past centuries. *Climate Dynamics* 37:35–51.

Nolan, T., and Crisanti, A. 2017. Using gene drives to limit the spread of malaria. *The Scientist*, January.

Passmore, D. 2016. Greenpeace again caught using misleading photos in Great Barrier Reef campaign. *The Courier Mail*, March 26.

Shewry, P. 2016. Editorial. *Journal of Cereal Science* 70:A1.

Teicholz, N. 2014. *The Big Fat Surprise*. Simon & Schuster, New York.

Watson, J.D., and Crick, F.H.C. 1953. A structure for deoxyribonucleic acid. *Nature* 171:737–738.

Willett, W.C., and Ascherio, A. 1994. Trans fatty acids: Are the effects only marginal? *American Journal of Public Health* 84(5):722–724.

chapter eight

Implementing scientific thinking and critical analysis

When people listen to political discourse or read articles in the press, they may respond in different ways. Those who are of the same political persuasion as the speaker or writer might agree with most of what is heard or read. At the other extreme, some who listen to or read the same thing may consider it to be an endless spiel of glib language, spin, and empty rhetoric. What are we to make of this apparent contradiction? The spectrum extending from blind acceptance to rejection gives a scale to measure the degree of critical thinking. In countries that are predominantly democratic, elected representatives align themselves with political parties. They are then committed to defend the policies of their party and to criticize the policies of opposing parties. When issues are debated, each member is obliged to toe the party line to a certain extent. There are limits to the scope for examining each issue on its merits. Scientists, on the other hand, should be detached from any particular line of thinking. Their aim is to arrive at the truth. They therefore have complete freedom to examine the issue. It may happen that they alter their opinion and support a view that they had previously opposed should new evidence present itself. Here, we are discussing a true scientist who is totally honest and not influenced by external factors such as, for example, sources of funding for the research.

What we see then is a sharp distinction between true scientific thinking and political thinking. Those who use scientific thinking search for truth, are prepared to listen to those with opposing viewpoints, and are open to changing their opinion if the arguments presented are convincing. They are concerned with long-term solutions to problems in order to improve the well-being and quality of life for everyone. Unfortunately, these are not the qualities that necessarily lead to a successful political career. Changing of one's opinion is often seen as a weakness in a politician. Flip-flop is a dirty word that can be used to denigrate. Success in politics requires a steadfast pursuit of a certain policy direction, even though this may at times not be in the best interests of the public. A skillful politician only has to convince a large proportion of the public that what is being proposed is in their best interest.

How then can we try to remedy this apparent paradox? One way is to encourage more scientists to enter public life. Some of the reasons why scientists are not attracted to politics were discussed in Chapter 4. It might be expected that there would be a realization that more scientific thinking is needed to help solve some of the current problems. It would be valuable to have a greater input from the analytical thinking that scientists can offer. In the absence of this, what is more worrying is that decisions about science are being made by nonscientists without an appreciable input from scientists. The disastrous consequences of this were highlighted in Chapter 4 in regard to the erosion of a once-renowned research organization. Fortunately, some programs are currently being developed to facilitate the entry of scientists into political careers (Todd, 2013).

Research scientists in federal parliament

Prior to 2016, Australia had one research scientist, Dr. Dennis Jensen, in its Federal House of Representatives. He carried the hopes of the scientific community on his shoulders. His task was not made easy as was demonstrated when he gave his maiden speech to the parliament. To expand on this, some background needs to be given. In 2010, the Australian federal election produced a stalemate with no clear winner for either of the two major parties: the Labor Party and the Liberal/National Coalition. The Labor Party succeeded in obtaining a majority by negotiating with the Greens Party and two independents to allow them to form a government. One of the issues that was negotiated was the introduction of what has been ambiguously referred to as a "carbon tax." This is essentially a tax calculated on the emission of carbon dioxide when hydrocarbon fuels (coal, petroleum, natural gas) are burned in industrial processes. Carbon dioxide is one of the "greenhouse gases" that becomes trapped in the atmosphere. It forms a blanket that reduces the loss of heat and this should in theory lead to warming of the planet. Debate about the magnitude of this effect is what is at the heart of the controversy about anthropogenic global warming. There has been a large proportion of the population that has been inclined to be convinced that the degree of warming could be large. The consequences would then be catastrophic effects on the earth's climate, causing such things as destruction of crops, rises in sea levels, and increasing intensities of major weather events like hurricanes.

Public opinion has been greatly influenced by reports from the Intergovernmental Panel on Climate Change (IPCC). The deliberations of this panel, comprising scientists and bureaucrats from many nations, have been discussed in Chapter 4. The message given by the panel has been that global warming will become a serious problem and that nations need to take drastic measures to counteract it. However, there are large numbers of scientists who question the conclusions of the IPCC and are critical of its evidence. These are often called "climate change deniers," although since they are mainly scientists, it would be more correct to call them "skeptics." Skepticism is an essential quality of a scientist. It is a positive attribute, not to be confused with terms such as cynicism, dogmatism, or obscurantism. The skeptics have been attacked for their lack of sound evidence as measured by peer-reviewed publications. It does appear that the case for anthropogenic global warming has been supported by a much greater number of peer-reviewed publications. There are, however, other factors that should be considered. First, research on the topic is mainly funded by bodies whose aim is to confirm that there is significant global warming. Thus, more publications would be expected. Second, because global warming has become the politically correct stance, it is more difficult for skeptics to get their papers accepted for publication in peer-reviewed journals. It is therefore not a level playing field. Those who submit papers that follow the more politically correct view tend to be shepherded through the system, whereas those who present the contrary view find it much more difficult to get their submissions accepted.

Against this background, Jensen, the only PhD research scientist in the Australian parliament for a period prior to 2016, gave his maiden speech. At that time, he was a member of the opposition party. The speech was a balanced presentation in which he succinctly discussed evidence for the contrary view to the politically correct one. That is, he presented evidence that challenged some of the claims that had been put forward to support the anthropogenic global warming case. As a consequence, he asserted that there was no justifiable scientific reason for introducing a carbon tax with its inherent damage to the national economy. During the speech, he asked to table the relevant peer-reviewed publications. His request was denied by the government. Although the attempt to table the documents was unsuccessful, the refusal illustrates an important point that everyone needs to think about. Here was the only research scientist in the parliament. It might be expected that his insight would be valuable to the debate since the question of anthropogenic global warming is predominantly a scientific issue. Tabling of the documents would have refuted the argument that there was a dearth of peer-reviewed papers questioning the prevailing view and would have made a positive contribution to the debate. The action highlights the difference between science and politics. The aim of science is to welcome dissenting viewpoints so as to stimulate the debate and try to approach the truth. The political bent is often to stifle attempts to present viewpoints that clash with the party line. It is not difficult for rational people to see which approach is more likely to arrive at a truer and more satisfactory outcome.

In the 2016 Australian federal election, Jensen lost his seat, so his input disappeared from parliament. The views he brought were carried on for a short time as a result of the election to the senate of Malcolm Roberts, a member of a minority party who has a technical background in engineering and has made an intensive study of the climate change issue.

Unfortunately, Roberts was made to leave parliament because of a technicality associated with his citizenship, leaving a gap in knowledge on this controversial topic.

How to introduce more scientific thinking

How can we make public decision making more based on truth than on political expediency? One way, as suggested in the previous section, is to elect more representatives who either are scientists or who are able to think scientifically. This would certainly be a step in the right direction, but this alone is not certain to work. Very often, if one party tries to tell the truth to an electorate and offers solutions based on this, it will not necessarily gain substantial support from the people. This is because a large proportion of the public does not want to hear the truth. They prefer to be told things that they like to hear and the truth is not always pleasant to hear. That is why many politicians who are successful will feed the electorate with things they like. The only way to bring about changes to the status quo is to develop a higher level of critical thinking in the voting public to ensure that political candidates are better held to account. This can only be done through education.

Changes needed in the education system

It is vital that, in the education system, standards of literacy and numeracy are maintained and, preferably, improved. There are, however, areas that seem to be neglected in schools. For example, simple basic hygiene does not seem to be given enough emphasis. Thus, we see people coughing and sneezing around us without taking the simple precaution of covering their faces with a handkerchief, while others do not wash their hands after touching surfaces that could be contaminated. Many people do not appear to understand the elementary science behind how infections are spread by transmission of germs through the air and by contact with surfaces. As a result, much unnecessary illness spreads through communities with consequent suffering and loss of productivity.

Another area that does not appear to be given sufficient attention in school education is nutrition. The lack of balanced nutritional diets for many people is the origin of much illness and disease. The preparation of nutritious meals is given little attention outside of home economics courses. Further, the crippling habits of smoking, drugs, and alcohol abuse are a heavy drain on society, as well as individual's lives. As a result, the cost of medical treatment places a high impost on society, much of which could be reduced by greater emphasis on good health choices and preventive medicine. Money management is another skill that seems to be given little attention, even though it is of crucial importance for almost everyone. The inability of many people to manage credit and to incur debt from which they become imprisoned, could be considerably reduced if more instruction about financial management were to be incorporated in education.

Perhaps one of the most glaring omissions from school curricula is that of the development of critical thinking. Before giving some thoughts on this important topic, some general points will be mentioned. The usual mind-set that people have of school education is that of students seated at desks in classrooms for hours while teachers instruct them on various subjects. At the end of courses, students are quizzed on what they have learned. This seems to be a reasonable and logical system for instruction and evaluation of students' performance. Most would agree that learning and testing are essential components of education. However, we should always be questioning the best way to achieve it. The majority of young people of school age are full of energy and their confinement to sitting for long periods is not natural. It tends to discourage the interest and engagement necessary for stimulating learning. Although there are recess and sports periods, the times spent in these activities are often small compared to the times when they are inactive. John Medina, a brain development scientist at the University of Washington, has suggested that the design of most classrooms is less than ideal. He proposes that, instead of desks, he would fill classrooms with treadmills or other gymnasium equipment (Medina, 2014). The idea is that better learning would be achieved by periods of aerobic exercise interspersed with focused learning periods. Effective learning demands intense mental effort, which, in turn, requires good physical health. The path to good physical health of young people is by exercise and sport, and not through a sedentary lifestyle as practiced in many school environments. Medina has suggested a more scientific approach to designing school curricula. Instead of relying on a committee to arbitrarily come up with a curriculum and inflicting it on students, different combinations of exercise and teaching could be tested to determine the best system for learning.

John T. Gatto (2005), in his provocative book *Dumbing Us Down: The Hidden Curriculum of Compulsory Education*, presents the view that public education does little except teach young people to follow orders. His view is that conventional schooling drives out the natural curiosity that children have. He believes that the system should be changed to one that empowers people to take control so that they are motivated to become lifetime learners.

The role of wonder in childhood education

From an early age, children feel wonder. It is vital that this sense of wonder be nurtured, and this can be done by exposing children to phenomena that instill wonder. In big cities, this is becoming increasingly difficult to achieve and needs a good deal of planning. For example, the experience of observing a clear, star-studded sky on a moonless night away from artificial lights is one that is bound to awaken awe and wonder. Other examples are to be found in nature: waves breaking on a deserted shore, the splendor of giant trees in a forest, rainbows, and the magnificent forms and colors of flowers and orchids. Many children, especially city dwellers, are often starved of such experiences. Children who experience the beauty of nature are likely to be stimulated to ask questions about what they have witnessed. These questions can then be brought up in the classroom setting. The important thing is for children to question. It may not be so important to provide satisfactory answers to their questions. In fact, it is probably more important to give answers that, instead of satisfying curiosity, stimulate further, and perhaps, deeper questions. If children become accustomed to questioning, this will provide a basis for later critical thinking. On the other hand, if they are fed dogma that they are persuaded to accept without challenge, the capacity to question is lost and they become puppets who are unable to think for themselves. This, to some extent, was the situation prior to the Enlightenment. During the Enlightenment, people began to reject the acceptance of dogma and to question widely held beliefs. This is what has led to the subsequent spectacular advances of civilization. Where submission to dogma occurs, no advances are possible and the culture stagnates.

Teaching with magic

The conditions that are employed for teaching in schools are critical for the success of learning. If these conditions are such as to instill interest and excitement, they will go a long way toward promoting motivation and engagement of students. Should the conditions fail to do this, the result can easily be boredom and disengagement. One way of preserving the sense of wonder and maintaining the engagement is the use of magic (Hatziapostolou, 2013). Magic includes many different puzzling effects such as making things vanish and predicting future events. Incorporation of these effects in classes, perhaps accompanied by humor, can break the monotony and maintain interest. The deception used by magicians takes advantage of assumptions that we usually make in everyday life and can teach us to question them. If students are asked to think about some magic tricks and try to explain them, this may help them to discard prejudices and apply critical thinking. This, together with performing some simple tricks themselves, helps to develop their confidence and self-esteem, which can be beneficial for coping with other problems they will face in life. Of course, for successful application, the teacher needs to have become proficient in some magical tricks, which requires a considerable amount of effort and practice to make them effective. Those who continually question are ones whose minds are likely to develop more during their lives. Their search for knowledge keeps them mentally stimulated and they are more likely to make positive contributions to society. In contrast, those who swallow dogma and adopt fixed viewpoints in their early years will have minds of stunted growth and have little to contribute to a better society.

Creativity

The introduction of wonder and magic into the syllabus might be accompanied by accounts of human achievements likely to stimulate young minds. They could include stories of genius such as those of the great composers, artists, and scientists. This would lead naturally into discussions of creativity. Frequently, some of the great insights are made by individuals who attain special states of consciousness. Rather than expanding on this, two quotations will be offered: one from a great scientist and one from a great composer. Albert Einstein stated, "The finest emotion of which we are capable is the mystic emotion. Herein lies the germ of all art and all true science. Anyone to whom the feeling is alien who is no longer capable of wonderment and lives in a state of fear is a dead man." Johannes Brahms said, "When I feel the urge, I begin by appealing directly to my Maker and I first ask him the three most important questions pertaining to our life here in this world—whence, wherefore, whither? I immediately feel vibrations that thrill my whole being. These are the spirit illuminating the soul-power within, and in this exalted state, I see clearly what is obscure in my ordinary moods: then I feel capable of drawing inspiration from above, as Beethoven did."

Introducing critical analysis

Although critical thinking should be encouraged from the earliest age, critical analysis may be best introduced as part of the curriculum during secondary school. This is not to suggest that it is not being done in some schools, but it does not appear to be as widely applied as it should. A good way to do this is to present the class with excerpts from newspaper articles or audio recordings of press statements from politicians or media personnel. The class is then invited to analyze what is read or heard and to offer commentary. Students should be encouraged to identify statements that contain falsehoods, distortions, or exaggerations. Their input need not be wholly negative. They should also be asked to point out examples of truth and fairness. A summary of the responses could then be sent to the sources of the statements, whether they be politicians, media representatives, or advertising agents. Such an exercise could have two positive outcomes. First, it could serve as good practical training in critical analysis for students. Second, by alerting the persons responsible that their statements are being monitored, they are likely to take greater care in what they say or write.

Some simple examples

The initial classes could present examples to illustrate faulty reasoning. Let us look at some obvious ones of the sort that appear regularly and see how they might be analyzed.

1. "There is compelling evidence that ..."

   If there is compelling evidence, many people will assume that it must be true. The critical approach is to ask "What is the compelling evidence?" It may be that the writer/speaker does not know what the evidence is and is merely making the statement to try to enforce a viewpoint. If evidence can be provided, then we need to examine whether it really is compelling enough to justify the conclusion that is being suggested.

2. "Tests have shown ..." or, even more convincing, "University tests have shown ..."

   If tests have shown, then it must be true. Or must it? Let us see the results of the tests. When we analyze them, do they really show what the writer claims or is there substantial doubt?

3. "I am reliably informed that ..."

   Many who read or hear this statement will assume that it must be true. But what is the source of the information and what is the criterion for describing it as reliable? Perhaps it is just a bluff to convince the audience that it is true.

4. "Clearly, this shows that ...."

   This type of expression is often used to persuade the audience. If it were clear, there should be no need to emphasize its clarity.

5. "This is the worst (government, prime minister, treasurer) the country has ever had."

   This is typical of the exaggerated statements that are bandied about, often without being challenged. How is the grading from best to worst objectively measured? What it comes down to may be just a subjective opinion, and some may agree and some may disagree.

6. "This statement is the stupidest thing I have ever heard."

   Such a declaration should be easily scotched. It is an example of personally attacking the one who made the statement without entering into a rational debate about the issue. However, this type of ploy is frequently used without being challenged.

7. "The people reject this idea."

   Really! All the people? This is a commonly used ploy and is often not challenged. Was this statement based on a survey of all the people or was it just the view of the person without any basis for extending it to everyone?

8. "God has said: You must kill all (Calithumpians)!"

   This is possibly the most difficult type of statement that critical thinkers have to deal with. One might ask, "Where is this stated?" Perhaps it is stated in a holy book. But, in another part of the book, it may say the opposite. For rational people, this could be a basis for debate. However, if we are dealing with people who are not rational, it may be impossible to have a debate. They may say, "We have a guru who instructs us and we believe our guru because he tells us the word of God, who is the ultimate authority." It is impossible to debate someone who has an entrenched dogmatic view as no amount of reason can ever change it.

9. "It is always the case that …" or "It is never the case that …"

   Always and never are absolute words. Usually, such a claim is an exaggeration and the argument can be logically refuted if exceptions can be provided.

Deflection of an issue

A common trick that is used to confound an argument is to deflect the issue, in much the same way as magicians deflect the attention of observers by their manipulations. Sometimes it may not even be done deliberately, but nevertheless it can be effective. Let us look at a scenario to illustrate how it works, based on a recent television debate. The names of the participants will not be given as they are not needed. The debate concerns a decision made by a political leader. One participant supports the decision and provides some facts to back it up. The opponent does not directly address the issue but instead attacks the integrity of the leader. To support this position, the opponent declares that hundreds of people turned out to protest against the leader and the decision.

The interviewer failed to pick up on two points. The first was that the issue was deflected by attacking the leader rather than criticizing the decision, a tactic often referred to as *ad hominem* or sometimes as playing the man and not the ball. Second, the fact that hundreds of people turned out to protest is irrelevant to the discussion. Hundreds, and possibly thousands, did not turn out to protest. On many political issues, the community is divided, frequently in roughly equal numbers. The fact that a large number protested means nothing. This is not to suggest that protests should not occur, simply that it is not an effective debating point in this argument.

Putting labels on opponents

A tactic that is being used more frequently, especially in political debate, is to put a label on an opponent (e.g., a racist, a homophobe, or far right/far left). This is aimed at silencing the opponent because no one likes to be tainted with a label that might be perceived as negative. The effectiveness of the ploy may show up in democratic elections where the result of the election turns out to be opposite to what was predicted by prepolling. People may be reluctant to state their opinions for fear of being branded with negative labels and prefer to reserve their views for expression in the shelter of the secret ballot box. It might be thought that it would be simple to point out that labeling opponents is not addressing the issue being debated, but such is the lack of critical thinking in many circles that this tactic is not called out.

The straw man argument

The straw man argument ploy aims to create the illusion of having refuted an opponent's proposition through a covert replacement of it with a different proposition and then refuting the false argument instead of the opponent's proposition (hence, knocking down a straw man). If the opponent is not alert, this can often be an effective technique to obfuscate the issue by tricking the opponent to go on the defensive.

Could have or might have

Another ploy that may seem obvious but for which many allow themselves to be misled is to resort to speculation. In order to cast doubt on someone's actions, it is speculated that this person could have or might have performed a dishonest or nefarious act. There is no evidence for it. However, by suggesting the possibility may often be sufficient to persuade those who are not trained in critical thinking to believe that it occurred.

The prism of ideology

When light passes through the interface between one medium and another, it is deflected. The effect is seen when white light passes through a glass prism. Its components are separated according to their wavelengths and we see a spectrum of colors from red to violet. This is a good analogy for the way humans process the information they receive. The mind acts as a prism in which facts may come straight through or be deviated by the mind-set. Thus, some individuals interpret facts so that they are consistent with their ideological beliefs. They may not be aware that this is happening and see their interpretation of the facts as simply being the truth. Many commentators in the media consistently interpret the news from a particular ideological position. They are not aware that their commentary is biased. Because of the ideological prism that operates in their minds, they believe that they are presenting the facts truthfully.

The influence of cultural background on the capacity for scientific thinking

In my interactions with research students in science from different countries and cultures, I have observed a striking difference between those who have lived in a relatively free society and those who have had the misfortune to be brought up in a society where ideology has been imposed. The ideology may stem from political or religious origins. In the free society, questioning is encouraged. People from this culture tend to be imaginative and, as a result, creative. They are open to new and profound ideas. Those from societies where ideology is imposed, show a lack of the flair that is needed to create substantial scientific advances. It may happen that some of these people do not agree with the ideology. However, in a strange sort of way, the culture appears to permeate the society to such an extent that their imaginative capacity is debilitated. For those who become researchers, their research often lacks the spark of imagination and their contributions tend to be pedestrian.

References

Gatto, J.T. 2005. *Dumbing Us Down: The Hidden Curriculum of Compulsory Education*. New Society Publishers, Gabriola Island, Canada.

Hatziapostolou, T. 2013.Learning begins with wonder: Engaging students through teaching with magic. In *Proceedings of the 7th International Technology, Education and Development Conference*, Valencia, Spain, March.

Medina, J. 2014. *Brain Rules*. Scribe Publications, Carlton North, Australia.

Todd, M. 2013. As an experiment, let's put more scientists in Congress. *Pacific Standard*, December 21.

chapter nine

Bringing it together

It seems that it would be useful to summarize some of the most important issues that have been covered in the previous chapters. The intention is to make these issues interconnected rather than remaining a series of separate topics, so as to unite them into a common theme.

"The need for critical thinking and the scientific method" is in the book's title. The rationale behind the title is that a lot of the problems that exist in the world today arise because many of its people are not able to think in a critical and unbiased way, and do not understand the scientific method and its contribution to thinking. As a result, they are easily indoctrinated by those who would use the situation to impose their agendas. These agendas are often not in the best interests of the world and its people. For those who are able to think critically, the Enlightenment period changed civilization for the better. Those who base their thinking on inflexible dogma do not share this view. Thus, we have two opposing forces that will determine the future of our civilization. The progress made during the Enlightenment period is not guaranteed to continue. It can only continue if there is a general awareness of the need to nurture the use of reason and imaginative thinking that contributed to this progress.

Scientific/critical thinking

First, what is critical thinking? There is a vast literature on the topic and there are a number of ways of defining it. Critical thinking means taking charge of our minds so that we can take charge of how we live. Tama (1989) called it "a way of reasoning that demands adequate support for one's beliefs and an unwillingness to be persuaded unless support is forthcoming." The National Council of Teachers of English (NCTE) Committee on Critical Thinking and the Language Arts defines critical thinking as a "process which stresses an attitude of suspended judgment, incorporates logical inquiry and problem solving, and leads to an evaluative decision or action." Ennis (1987) suggests that "critical thinking is reasonable reflective thinking that is focused on deciding to believe or do."

Critical thinking is a foundation of science and is an important component of what has led to the spectacular increase in understanding of the physical world that has occurred, especially in the past few centuries. Scientific thinking is critical thinking but also involves understanding of the philosophy of the scientific method, the elimination of bias, and the capacity to consider the viewpoints of others that may be different to your own. It is usually impossible to eliminate bias completely, but it should be a goal for all to strive for. The willingness to listen to the opinions of others is not just an altruistic gesture. Research scientists often profit from the insight they derive from debating issues and hearing the input of others with opposing views. The same is true for debates in general. Opposing viewpoints can open new windows and cause people to rethink the issues. It may cause them to change or, at least, modify their opinions. On the other hand, it may help them to see the fallacies in alternative arguments and thus make them feel more confident in what they have believed. In either case, it proves beneficial.

When we discuss scientific thinking, we need to understand what is meant by "scientific." There is a good deal of misunderstanding among the general community of what is meant by the scientific method. Many believe that the term scientific is synonymous with empirical. Certainly, careful observation and accurate measurements are key components of the scientific method. However, it is widely believed that if an observation is made that is consistent with a previous independent observation, this can be taken as a further confirmation of a theory. This is not so, as was made clear by Karl Popper (2002). To explain this, let us look at a topical example. If we observe that melting of ice in the arctic region is increasing, this suggests that the planet may be warming, although it does not necessarily mean that the effect is due to human activity. If a further observation is made that the glaciers are disappearing in another part of the world, this is also consistent with warming. But that is all it is. It simply means that each successive observation can be interpreted as being consistent with the theory of anthropogenic global warming. They cannot be used, in the strict scientific sense, to count as additional confirmations of the theory. In the case of melting glaciers, this could be due to other causes (e.g., natural variation in climate).

The scientific/critical versus the dogmatic approach

The search for observations to find confirmations of a theory is a characteristic of the dogmatic approach. The dogmatic approach can serve a useful role in science. It can provide information to be used in the formation of hypotheses. It can be thought of as a precursor to the scientific method. The true scientific method or critical approach begins with a hypothesis that attempts to explain previously apparently unconnected observations. The hypothesis is the creative step in the scientific method. It requires imaginative thinking. Once a hypothesis is proposed, experiments are designed to severely test it. This is the logical or deductive part of the method. Thus, the true scientific method, as proposed by Popper, is the hypothetico-deductive method. The inductive method, which relies on repeated observations that are used in an attempt to arrive at a generalization, is not a valid method for discovering knowledge, as has been shown by Popper. The transition from using a dogmatic approach (induction) to using a critical approach (hypothesis-deduction) represents an evolution in scientific thinking. Whereas the dogmatic approach tries to confirm a theory, the critical approach designs experiments to try to refute the theory. One essential difference between the two methods is that the inductive method tries to proceed from observation to theory, whereas the hypothetico-deductive method tries to proceed from theory to observation. The requirement of a good hypothesis is that it is refutable, that is, testable and falsifiable. A hypothesis that is not refutable is not scientific. If the hypothesis is refuted, it is rejected and a search commenced for a new one. The new hypothesis makes use of what has been learned from the failed hypothesis. Thus, science progresses by a trial-and-error procedure in which we learn from our mistakes. A weakness in the anthropogenic theory of global warming is that it does not appear to have been subjected to a refutable hypothesis. This does not mean that the theory is wrong, simply that it has not been scientifically tested. Thus, the assertion that the science is "settled," which is often made by many, including those who call themselves scientists, is false. It is based on the misconception that each new observation that is consistent with a theory can be considered as an additional confirmation.

The diminished role of scientists

When we consider that understanding of the physical world and improved living conditions of humans have resulted mainly from science, the absence of scientists from public office is striking. Even more remarkable than this is that many decisions about scientific matters are not being made by scientists. This has manifested itself in the displacement of outstanding scientists at the head of research organizations and their replacement by managers who are nonscientists. The removal of scientists from leadership positions in their field is not going to be helpful for inspiring the general public to inject the methods of science into their thinking.

Recent decades have seen a shift in the hierarchical status of employees of scientific organizations. Managers have assumed high positions while active scientists have seemingly been relegated to positions of lower status. Salaries of those in management positions relative to those in scientific research have increased to reflect this perceived change in balance.

Decisions about how research is organized are being increasingly made by managers who are often nonscientists. Furthermore, criteria for evaluating the performance of scientists are often designed by administrators. Managers are concerned with deciding objectives and mapping out the steps and timelines needed to achieve them. The organization of research is therefore reduced to something akin to what occurs in a factory production line. This approach is completely at odds with how scientific knowledge is advanced. The essential difference is that the managerial approach begins with sharply defined objectives and follows a more or less blinkered course to target the objectives. This means that nothing more can be accomplished than what was originally planned. In contrast, the true scientific approach also may begin with objectives, but these are flexible and can lead to exploration of ideas that may unlock discoveries not imagined at the beginning of the research. The scientific approach differs from the managerial approach in that it opens the way for imaginative thinking. It is what distinguishes science that is great from that which is pedestrian. The contrast between the two approaches, as mentioned earlier, is captured by a quotation of Albert Einstein: "Imagination is more important than knowledge for knowledge is limited to all we know and understand, while imagination embraces the entire world and all we ever will know and understand."

The effects on scientific excellence

The result of the encroachment of managerialism into the organization of scientific research has caused a lowering of quality. This is shown by an explosion in low-quality published papers in scientific journals (Bauerlein et al., 2010). The pressure put on scientists to achieve mundane targets, to satisfy timelines, to spend much of their valuable time in chasing grants, and to be prolific publishers of papers inevitably leads to the erosion of quality. The emphasis on quick solutions, often in the form of short-term consultancies, threatens the employment prospects of scientists who have dedicated their careers to acquiring a deep knowledge base that befits them to undertake long-term research of a fundamental nature. An unhealthy work environment forms in which scientists have to aggressively compete for funding, some of which is apportioned to politically correct issues.

It is not surprising that the changes that have been imposed on scientific endeavor have caused dissatisfaction and a lowering of morale among many scientists. One example where this has been manifested in a research organization (the Commonwealth Scientific and Industrial Research Organisation [CSIRO] of Australia) was discussed in Chapter 4. An Internet blog, Victims of CSIRO, was set up to allow scientists to air their grievances.

How scientific organizations can be infiltrated by destroyers

The formation of unhealthy work environments often facilitates the advent of individuals who prosper from these conditions and utilize them to advance their careers at the expense of the organization and some of its employees. Some personality traits were discussed in Chapter 4, but the one that causes the greatest damage is that of the psychopath. Most of these people are indistinguishable and remain hidden in the crowd. The psychopathic trait is practically not understood by the general public. Psychopaths form only a small percentage (perhaps about 2 percent) of the population, but the damage that they cause can be much greater than suggested by this proportion. Furthermore, it is found that their presence tends to be more frequent in management positions than in the general population.

It has been suggested that many of the failures of corporations in recent times can be ascribed to the presence of psychopaths. They infiltrate organizations and destroy them from within. Then, after the corporation disappears, they move on without incurring any blame and subsequently often prosper. Scientific institutions are not immune to the destructive influences of psychopaths. One of the characteristics of psychopaths is their pathological lying. As mentioned in Chapter 4, an Appeals Tribunal in a compensation claim by a former CSIRO employee found no less than 128 false or misleading statements by two CSIRO employees, one of them being a senior executive. It has not been shown if this was psychopathic, but, in any case, there seems to be a need to better recognize the dangers posed by psychopaths in scientific organizations and how to identify and nullify them.

The pros and cons of democracies

As Winston Churchill famously said, "Democracy is the worst form of government except for all the rest." What this seems to imply is that, although a democracy can be the best form of government, it does not mean that it is free of faults. One obvious fault is that an individual who gives little thought to issues has the same vote as a person who becomes informed and thinks deeply about them. Certainly, if we look at different forms of government in the past, democracy stands out as the one that gives the option of a voice to all and promotes the most individual freedom.

The ideal democracy is one that is made up of people who are capable of thinking critically. This enables those of good faith to discuss issues with minds that are open to considering alternative viewpoints. How a democracy can fail is where a society has a dearth of critical thinkers. Such a situation allows unscrupulous individuals to manipulate people and impose views that may well be false.

How can we determine if a proposed new law is better than the law it replaces?

Suppose a law that changes a nation's constitution is proposed, requiring a public vote. In a democratic system, if a majority votes in favor of the new law, then its introduction is legislated. Will this mean that the new law is better than the old one? Perhaps, but not necessarily. All that can be concluded with certainty is that more voted for the new law than voted against it. That is how a democratic system works. How then can we know if the new law is better? There is no simple answer. The best that can be done is to have a public debate on the issue prior to the vote. How close the debate comes to arriving at the better alternative depends on the quality of debate. If the debate proceeds with dogmatic arguments, trading of insults between protagonists, and political point scoring, then chances of arriving at the best result will be diminished. If, however, the issue is debated by rational protagonists who base their arguments on critical thinking and respectful consideration of all viewpoints, the probability of achieving an optimum result will be enhanced.

It's the electorate, stupid!

When it comes to elections for a governing body, the final result may depend not necessarily on the merits of policies proposed by opposing parties or candidates but on the makeup of the electorate with regard to their capacity for critical thinking. If all who form the electorate are critical thinkers, the issues are more likely to be logically discussed and the election more likely to result in an outcome that is optimum for the nation. Political candidates who offer themselves for election then need to present policies that stand up to critical scrutiny. If not, they will either fail to gain support or will have to change their policies so as to withstand critical examination.

How does the makeup of a population determine its government? The ideal of having a community made up exclusively of critical/scientific thinkers is just that—an ideal. What do we see in the real world? In democratic nations, the composition of constituents encompass a spectrum of individuals in regard to thinkers. At one end of the spectrum are those who are genuine critical thinkers who are unbiased and open to different opinions. At the opposite end are those who are not prepared to ponder issues and usually form dogmatic opinions based on a few experiences and hearing catchphrases or slogans. In between the two extremes there is a gradation in capacity for critical thinking. Of course, we cannot put numbers on the different groups. Humans are more complex than that. However, on the basis of surveys (and political parties carry these out), it is possible to get an idea of the composition of the electorate in terms of critical thinking capacity.

Suppose, from a survey, we find that 10 percent of the population can be classified as genuine critical thinkers, whereas 30 percent give little or no thought to political issues. Candidates for election must try to persuade as many people as they can to give them their votes. Therefore, where will they concentrate their efforts? The 10 percent is more likely to include people who give consideration to the individual or party that promises most in terms of the national interest. They will tend to favor long-term solutions. Those in the 30-percent group are likely to give greatest consideration to how they will benefit personally and the national interest will be secondary. They will be more attracted to short-term solutions of problems.

Another factor is whether the election is based on compulsory voting or not. In some countries, such as Australia, voting is compulsory. What this means is that on polling day, all the people who are on the electoral roll need to attend a polling station and have their names crossed off or submit a postal vote, which requires prior authorization. If they do not, they will be made to pay a fine. Some of those on the electoral roll may decide not to cast their vote at the polling station so, in a sense, voting is not compulsory although having their names crossed off is needed to avoid paying a fine. It is controversial whether compulsory or noncompulsory voting is the better system. Those who oppose compulsory voting may say that freedom to vote is better because those who vote will comprise a higher proportion of responsible critical thinkers. An argument in favor of compulsory voting is that, in theory, it is more democratic and encourages all the people to give thought to the issues.

Most politicians are astute enough to realize that they will get more mileage by addressing those who give little thought to the issues (the 30 percent mentioned earlier). How will they convince this group to vote for them? It will not be by reasoned arguments because this group will not have the time or inclination to listen. What is needed is something simple and easy to assimilate such as a slogan. Many in the 30 percent will decide to vote for those who seem to offer them the most. If, for example, we suppose that many of the voters receive government entitlements, one candidate can say that his/her opponent belongs to a party that is likely to cut their benefits. They may not intend to but by suggesting this threat, this may be all that is needed to sway a lot of voters.

Another effective ploy is to focus on someone in the opposing party, preferably the leader. If you say that this is a bad person who does bad things, and repeat this message continuously, that may be all that is needed to convince a large number of people to vote against the party. It is much simpler and easier to fix your hate on a single person than to go through the painstaking process of perusing different policies and evaluating them. Every time the person's name is mentioned, a signal is given to start hating, similar to the case of Pavlov's dog. Thus, we often see politicians repeatedly mentioning a person's name in order to start the conditioning response.

The media

As mentioned earlier (Chapter 5), a society is fortunate if it has a free media. In a totalitarian system (which may be autocratic or theocratic), the news presented to the public is censored and serious criticisms of the government are blocked. The ideal media is one that is not only free but is unbiased. In practice, although a country can boast that it has a free media, this freedom can be nebulous. If the news presented is slanted toward one political ideology, it may turn out that this sort of press is not that much better than the one that exists in a totalitarian system. It is said that if you do not read newspapers, then you are uninformed, whereas if you do read newspapers, you are misinformed. This is rather cynical but, sometimes, it may have some truth. The advantage of a free press is that it allows, at least in theory, opposite opinions to be presented. Therefore, unlike a totalitarian press, the possibility exists of examining different viewpoints and subjecting them to analysis, discussion, and criticism. For example, in the United States, the Fox News TV network presents a predominantly conservative viewpoint, whereas the cable networks such as CNN and MSNBC slant toward a liberal or progressive viewpoint. A similar division exists in the print media. Responsible media outlets try to incorporate alternative opinions. Thus, when there is a debate about an issue, they try to interview people with alternative viewpoints.

Unfortunately, there are some media outlets that only present one view. Critical thinkers are able to recognize the bias. The problem is that a large section of the public is not trained to think critically. These people choose the media that they wish to listen to and, as a result, will often only hear one side of an argument and any dogmatic views that they hold will become further entrenched.

How public opinion can be influenced by the media

There are many ways in which a biased media outlet can persuade readers or listeners to come to agree with its agenda. We saw two examples in

Chapter 5 where the public has adopted a perception that has been relentlessly pushed by the media. A more balanced scrutiny of these issues shows that they are not necessarily black and white. For each, an opinion has been formed by a large majority of people throughout the world that is not based on a balanced examination of the facts but on acceptance of a persuasive campaign by a section of the media with a specific agenda. These two examples are not isolated ones. On almost every issue, the media influences opinions and, if an appreciable proportion of the public are not critical thinkers, may induce a certain belief to be universally accepted, a belief that could turn out to be false or, at least, not clear-cut.

Media outlets use a variety of methods to convince a gullible public. They do not need to indulge in reporting falsehoods. In fact, this may not be a useful way because untruths are relatively easy to expose. There are more effective methods. These may involve subtle massaging of the facts without resorting to blatant lies. Another strategy that is commonly used is to not report facts that might detract from the intended message. Those in the media develop skills in communicating that enable them to pull the wool over the eyes of a more innocent and trusting public.

Opinion writers

Newspapers employ opinion writers. They contribute articles regularly, perhaps daily. All have their particular biases, although some may keep them well hidden. Thus, one opinion writer can have an enormous influence because his/her contributions are read by a large number of readers. Fortunately, means of countering this one-sided influence have evolved in recent times. Some newspapers allow short comments on articles to be published below the article (providing they are polite), thus giving the readers a say. Similarly, talk-back radio is a recent positive innovation. Although the hosts of these programs may push their own views, the same as opinion writers, it gives members of the public an opportunity to present their views. This enabling of an exchange of ideas is a valuable component of a democratic system.

Dangers to the progress of science

The positive contribution that science has made to civilization cannot be taken for granted to always continue. Similarly, the injection of the philosophy of the scientific method into critical thinking cannot be assumed to continue to advance. Like freedom, the advance of science is not guaranteed and can only be sustained by eternal vigilance. There are forces that are trying to erode science and scientific thinking. Some of these may result from the best of intentions but, because of a lack of understanding, they are misguided. In some ways, it is good to have resistance. Noble goals cannot be easily achieved. They require pitting of the intellect against opposing forces.

The lack of imagination or real interest

Karl Popper outlined some of the dangers that science confronts and these were mentioned in Chapters 2 and 4. One is the lack of imagination or real interest. There have been times in our history when the spirit of inquiry has waned and, in some cases, has stagnated. Comfort and easy lifestyles are not conducive to achieving great things. Great accomplishments are usually achieved only through struggles. The waning of real interest in inquiry and imaginative thinking can have and have had a variety of causes. I would like to focus on one that I believe is presently contributing to this demise. It involves the influence that managerial policies are having on the quality of scientific output.

How managerialism erodes the standard of scientific research

As noted in previous chapters (Chapters 2 and 4), the replacement of scientists by managers in the hierarchy of scientific institutes is having the effect of stifling the imaginative thinking that is essential for scientific excellence. The planning of research projects has to some extent been taken from scientists and placed in the hands of managers who do not understand how science works. They believe that managerial principles can be applied to scientific research in a similar way that it is used to monitor the efficiency of mass production lines. Thus, goals are set and the procedures to achieve the goals are developed in terms of milestones, expected outcomes, and timelines to achieve them. The fallacy in this approach is, as stated earlier in the chapter, that only what has been planned can be achieved. In contrast, research that is not shackled by managerial control has unlimited scope for what might be achieved.

Is science progressing?

Karl Popper suggested that one of the requirements of science is its need to progress. Science progresses by researchers building on the advances made by previous workers. I have been editor-in-chief of a scientific journal for a number of years. The rate of submission of papers to the journal has been steadily increasing. Does this mean that significant new knowledge in the field is increasing proportionately? My impression from scrutinizing all submitted papers and copyediting all accepted papers is that it is not. How is the advance of scientific knowledge measured? There is no simple way of doing it, but there are indicators that can be used to monitor the quality of research. The first of these indicators is that there is an increasing proportion of submitted papers that are simply accounts of observations. They may contain tables of results and figures but very little if anything in the way of theoretical concepts. As pointed out by Popper, scientific progress does not mean a continuous accumulation of observations but modification and improvement in theoretical understanding. A second indicator relates to how well current research builds on the efforts of previous workers. Today, I see an increasing proportion of papers that ignore previous work that is relevant and important to the study being reported. This often results in the reinvention of the wheel. However, the problem is more than just duplication. The new wheel may be square instead of round. In other words, the science may not only not be advancing but may be regressing. This is the exact opposite to how science is supposed to work.

The reason why prior relevant research is ignored is not always clear. It may be that the authors have not carried out due diligence on the literature. In some cases, this may be because the literature on a topic may have become so vast (possibly due in part to an explosion in the pedestrian papers mentioned earlier) as to make it difficult to cover or that all the relevant literature cannot be accessed by the authors. If the latter is so, it is a serious fault of the system. Advances in science depend on previous work being available to all. It could also be that the authors do not have the fundamental knowledge to understand the work or do not have the perception to appreciate its significance. Whatever the reason, it represents a failure to uphold the standard needed for the science to progress.

If science is not advancing as it should, what is the cause?

The replacement of scientists by managers in the control of research in some scientific organizations has consequences, some of which have been alluded to. One is that a greater proportion of research is being directed to short-term objectives. There is thus correspondingly less long-term research, which is the type of fundamental research that advances theoretical knowledge. The objectives of research projects and the timelines to achieve them are being devised by managers. This managerial control of research tends to lead to pedestrian outcomes.

The performance evaluation of scientists is increasingly being made by administrators. One of the main criteria for evaluation is the number of papers that have been published in a certain time. This encourages scientists to give priority to turning out papers rather than genuinely trying to advance significant new knowledge. It is easier and quicker to collect results and present them in tables and figures than to have to travel up the dead-end roads that are needed to gain a deep understanding of a problem. The pressure being put on scientists to publish papers in order to survive in a competitive environment is causing them to cut corners. They may fail to carry out complete searches of the literature and to avoid the time-consuming work that is needed to make significant contributions. The obsession with publication rate at the expense of quality research is an example of the decrease in imaginative thinking and real interest that Popper warned about.

How scientific progress can be eroded by authoritarianism

Another danger to progress enunciated by Popper is the effect of authoritarianism on scientific endeavor. Some effects of this have been previously mentioned (see Chapter 2). There have been numerous obvious examples where scientific progress has been subjugated by authoritarian influence. The notion that the earth was not the center of the universe was resisted for a long time by religious authority. An example where political authority exerted an influence was the genetic theories of Lysenko in the Soviet Union. These were strongly supported by the ruling regime at the time and those who presented opposing views were banished. Since then, it has been shown to be flawed science.

The contrast between students and researchers who have grown up in societies that are subjected to political or religious authority and those in a relatively free society is notable. As a generalization, those who have lived under an authoritarian regime show a distinct lack of the imaginative spark and creative thinking that we see in free societies. The great advances in civilization achieved during the Enlightenment period were not made by those who based their thinking on dogma but by those who rejected dogma and questioned everything.

Since Popper's time, there has been another type of authoritarian influence imposed on science. This is managerialism, which was discussed in Chapters 3 and 4. Scientists are often placed under tight control, told what specific objectives they must pursue, and the timelines to achieve them. Of course, scientific research needs to have direction. What is important, however, is that working conditions for researchers are such as to allow them to develop their creativity and not be subjugated to the same control as those who work on a factory production line.

Flaws in the application of thinking to some current issues

We have looked at several controversial issues (Chapter 7) that require a scientific input into the critical thinking needed to analyze them. The topics considered have been anthropogenic global warming (often loosely referred to as climate change), dangers to the Great Barrier Reef, effects on health of saturated fats in the diet, the gluten-free food market and the topic of genetic engineering. For each of the topics, a proportion of the population has formed strong opinions, although the scientific basis for the opinions may be far from solid. Furthermore, acceptance of some of the arguments that have been made on these issues is questionable and putting them into practice could cause detrimental effects on human welfare.

For example, acceptance that human activity is changing the climate is having huge economic impacts. It has led to attempts to phase out coal, which has been a relatively cheap source of power. Admittedly, reduction of pollution from industrial emissions is an aim that should be supported. However, this is not the main reason being used to justify some of these policies. It is the theory that human activities are causing global warming and, if they are allowed to continue unabated, there will be catastrophic effects such as failure of crops and extreme weather events. The theory (or hypothesis) may be true but, as yet, it is doubtful if it has been corroborated scientifically. Nevertheless, many seem to have embraced it, very much like a religion, in which their beliefs are more based on faith than on reasoning. As we have seen, the theory has been based on an inductive approach. That is, that different independent observations are consistent with warming of the planet. However, the weakness scientifically is that each of the observations (e.g., rising sea levels, disappearance of arctic ice, melting of glaciers) cannot be taken as additional confirmations, as is usually claimed. They are simply observations that can be interpreted as being consistent with the theory. They could also be interpreted as natural variations of the climate and not necessarily due to human activity. In order to acquire evidence in favor of anthropogenic global warming, we would need to try to refute the theory, applying hypothesis-deduction, but very little of this approach appears to have been applied.

In the case of the saturated fat–heart disease hypothesis, this too has been based almost entirely on inductive evidence. Many studies have been carried out, but most have been epidemiological studies. This type of investigation can find correlations between saturated fat consumption and heart disease but is not able to demonstrate a cause-effect relationship. Nevertheless, in spite of the lack of a scientific foundation, this hypothesis has been widely promulgated and has had a profound effect on the diets that have been adopted in many countries over the past half-century. Evidence is also now starting to appear that casts doubt on the validity of the hypothesis or, at least, suggests that it should continue to be questioned. Furthermore, some studies are showing that adoption of a low-fat diet and replacement of fats by carbohydrates may be resulting in negative effects on health such as susceptibility to obesity and diabetes.

The use of induction rather than hypothesis-deduction is one failing of thinking on some controversial issues. Another common failing that manifests itself in the issues we have been considering is that of selection bias. Selection bias means that studies of an issue are chosen to fit a hypothesis and those that do not support it are ignored. For example, in the debate about the dangers to the Great Barrier Reef, we have, on one side, the environmentalists and, on the other side, the industrialists. Each of these groups will tend to push its respective agenda. Environmentalists are concerned with preserving a benign environment for protection of the reef and criticize those who might endanger the reef by such activities as mining and agriculture in the vicinity as well as shipping. Industrialists will try to play down the possibility of damage to the reef by their activities. It is good that there are inputs from each side. Where it can become problematical is when one side resorts to exaggeration or even false reporting to support their agendas. It is therefore essential that those who seek to understand the state of the reef and how it is being affected apply critical thinking to evaluate the information that is fed to them. It is vital to keep an open mind and not to form opinions based on uncritically accepting the information that is provided.

The growth in the gluten-free food industry again does not have a strong scientific basis. Certainly, there is a small proportion of the population for which the consumption of gluten has a detrimental effect on health. It includes those who suffer from celiac disease or other intolerances to gluten. These patients must avoid eating foods that contain gluten. On the other hand, the spectacular growth in the gluten-free food industry has occurred as a result of a campaign to persuade the general public that gluten is unhealthy. The avoidance of gluten-containing foods has become more of a fad, not based on scientific evidence. It has not been promoted by dietitians or medical specialists but predominantly by writers in popular magazines. A large section of the public has been hoodwinked by articles in the popular press and advertised by vested interests, the food manufacturers, who profit from the expanding sales of gluten-free foods. It is similar to the case of the fat-free diet in that evidence is now appearing that shows that elimination of foods containing gluten may be contributing to health dangers, specifically the aggravation of diabetes (Bodkin, 2017).

The controversy surrounding genetic engineering has also created a division between those who emphasize the potential benefits and those who point to the possible risks involved in flaunting nature. As with the other issues that have been discussed, it is good to have these opposing views. There are benefits and there are risks. The important thing is that each application of the new technology be honestly debated so that sound and unbiased decisions can ultimately be made for each.

In summary, the five controversial issues that were discussed are good examples of how the public can be influenced to take positions that may be wrong if they fail to apply scientific/critical thinking to them. Some criteria for bringing critical thinking to bear on issues were summarized in Chapter 5 based on the guidelines suggested by Elliot D. Cohen (2013). The appropriate way to consider issues is to weigh up the arguments on both sides and to suspend judgment while the issues remain the subjects of debate. In particular, we have seen that there are at least three main traps that people can fall into if they fail to apply the correct thinking. The first is that inductive reasoning cannot be accepted to establish a scientific basis. Thus, epidemiological studies, although valuable in some cases as pointers to gain understanding, do not justify firm conclusions. Second, the influence of vested interests needs to be taken into account, as they will often distort the issues. Third, the presence of selection bias, that is, considering certain facts and ignoring others that may not support a hypothesis, needs to be identified and closely scrutinized.

We have seen examples of how the absence of scientific thought can lead to unjustified conclusions and, as a result, poor decision making on important issues. There appears to be a need to introduce more scientific thinking into public discourse. There has been a dearth of scientists entering public life. Scientists do not necessarily make good politicians. However, they are needed to inform and advise decision makers of the scientific method and how it should be applied.

Education as the means to raise the standard of thinking

If we are to raise the level of thinking in society, it has to be done through education. Some of the topics that tend to be neglected in education were mentioned in Chapter 8. They included lack of instruction in hygiene, nutrition, and finance management. However, the main ones we are concerned with here are critical thinking and understanding of the scientific method. If we are to advance understanding in these attributes, it needs to begin with education. Critical thinking combined with instruction on the scientific method should be given more emphasis. If people are not trained in these skills, they can become puppets of those who would try to impose their agendas, including those in the media. The way to develop proficiency in any skill is to practice. Classes in critical thinking need to begin by providing practical exercises such as the few simple ones illustrated in Chapter 8. After learning to recognize some of the traps that one can fall for in these simple examples, more complex ones can be introduced.

Unfortunately, many people are subjected to dogma from an early age. Already, at preschool age, many have been indoctrinated with certain ideas that remain with them as they move through life. How can these influences be counteracted? This is a challenge. Some ideas have been suggested earlier (Chapter 8) such as introducing children to experience wonder and the use of magic in teaching. For example, if children (or any persons) have the opportunity to gaze at a star-studded sky on a moonless night free from artificial lights, this should instill wonder. The realization that they are situated on a rock that is orbiting a star that is one in billions in the Milky Way galaxy and outside of which there are billions of other galaxies should make them forego, at least for a time, the petty thoughts that have been instilled in them. The use of magic tricks is another way of making people shift their thoughts from the mundane to the sublime. It causes them to think and inquire. That is the sort of antidote needed to resist the submission to dogmatic viewpoints. An understanding of the philosophy of the true method for scientific enquiry is simultaneously needed in such a way as to stimulate the spirit of inquiry needed to replace some of the rote learning to which students may be subjected. Another consideration in teaching is to reduce the sedentary nature of receiving instruction by replacing it with a system that includes greater physical activity.

Dangers posed by psychopaths

Many students may have the misfortune to have their paths crossed by psychopaths. This is a danger of which most educators have no understanding. It is an issue that needs to be recognized if we are concerned about safeguarding students and future generations of scientists. Students learn to deal with conflict among peers. However, the skills they develop may be of no use if they should become the targets of psychopaths. This is a different type of human, characterized by not having a conscience. These people are able to wreck havoc on the lives of their victims without being exposed. As we have seen (Chapter 4), they can also destroy corporations and this includes scientific organizations. The education system needs to include, in syllabuses, invitations for psychiatrists with a knowledge of the psychopathic trait to visit schools to inform students about the danger. It seems strange that psychiatrists have not come forward to do this previously, as the destruction that is caused should have been obvious to them. The only way to counter the problem is to throw light on it. Psychopaths prosper by avoiding the light and operating secretly. This means that educators need to understand and be vigilant in recognizing the characteristics of psychopaths, how they operate, and how their influence can be thwarted. The way to do this is for everyone to be aware of how they target their victims and to combine to expose them so as to hinder their destructive behavior.

References

Bauerlein, M., Gad-el-Hak, M., Grody, W., McKelvey, B., and Trimble, S.W. 2010. We must stop the avalanche of low quality research. *The Chronicle of Higher Education*, June 13.

Bodkin, H. 2017. Is going gluten-free giving you diabetes? New study links diet with the disease. *The Daily Telegraph*, March 10.

Ennis, R. 1987. A taxonomy of critical thinking dispositions and abilities. In *Teaching Thinking Skills, Theory and Practice*, J. Baron and R. Sternberg, eds., 9–26. W.H. Freeman, New York.

Popper, K.R. 2002. *Conjectures and Refutations*. Routledge, London.

Tama, M.C. 1989. Critical thinking in every classroom. *Journal of Reading* 33(1):64–65.

chapter ten

Where will the future take us?

To answer the question posed by the chapter title, where the future will take us in relation to critical thinking and the scientific method, we can only resort to speculation. This is just as well. If we knew in detail what is going to happen, this might be depressing and it would certainly curb the striving for betterment that has characterized humankind's progress. Not knowing also means that the future is not predetermined so that there are no limits to what might be accomplished. The goals that we could achieve may be presently outside our capacity for imagination, just as some of the developments in the past century were not imagined in earlier times. It is not possible to predict future events with any certainty. All we can do is examine history and look at the trends that are currently occurring and where different paths may lead.

What do we learn from history?

What we learn from history is that life on earth appears to have evolved from unicellular organisms to more complex ones. Animals have evolved to adapt to the conditions. The advent of primates, culminating in *Homo sapiens* represents a paradigm shift in the evolutionary scale. The human species developed the mental capacity for reflective thought, something other species have not achieved to any degree. This capacity has allowed the human race to go beyond mere survival. It has enabled it to build civilizations with the accompanying advances in culture, the judiciary, art, architecture, literature, music, science, and the humanities. We can see in the evolutionary process, a progression toward superior life forms. Having acquired the capacity for reflective thought, there seems to be no limit to what might be achieved.

The theme in this book is that critical thinking combined with an understanding of the scientific method should enhance the quality of humankind. Based on how evolution has proceeded, always seemingly toward something better, we should be optimistic about the future. A society in which people become more proficient in critical thinking and have a greater understanding of the scientific method would be expected to be able to unite its members to form one that is more harmonious and just, and to provide benefits for future generations. The pointers from history are encouraging but what is uncertain is the time scale. In human history there have been times of great cultural progress (e.g., the Enlightenment period), but these have been interspersed with periods of stagnation or even regression. Although, over a long time scale, we have seen what we perceive as an advance toward a more enlightened and civilized society, the periods of cultural stagnation can be hundreds of years.

What trends do we see?

Will the period in the near future be one of advancement or regression? Let us look at some current trends and I will refer to Western countries such as the United States, United Kingdom, and Australia, for which I am most qualified to discuss. One trend is the increasing intolerance toward the views of others in public debate. This is contrary to how the scientific approach advocates how discussion should proceed. It has been pointed out in previous chapters how those who debate an issue are often divided into two groups who have irreconcilable views. To have opposing views is good because it stimulates discussion. If the participants keep open minds, it can be productive. When participants have entrenched opinions, are not open to considering changing or modifying them, and are intolerant to opposing viewpoints, this is when dogma takes over from scientific thinking. In these circumstances, there is a need to be aware of the paradox of tolerance as enunciated by Karl Popper (*Wikipedia*, n.d.): "Unlimited tolerance must lead to the disappearance of tolerance. If we extend unlimited tolerance even to those who are intolerant, if we are not prepared to defend a tolerant society against the onslaught of the intolerant, then the tolerant will be destroyed and tolerance with them. We should therefore claim, in the name of tolerance, the right not to tolerate the intolerant."

Presently, we are increasingly witnessing cases of people who wish to speak in public but are being prevented from speaking by opponents who do not accept that they should hold opinions different to their own. This may take the form of shouting down or it may involve intimidation and violence. Those who are trying to silence people of opposing views are predominantly from the left of politics. It hasn't always been so. Those of the left have at times been the champions of human rights while those on the right have been the despots. Many put Stalin on the extreme left and Hitler on the extreme right, but there is debate about where each fits on the political spectrum. It has been suggested that a horseshoe represents the political spectrum better than a left–right linear continuum. Although proponents of the left and right do not agree on this, the model does have some credence in that extremes of the two ideologies approach each other by having a shared trait of totalitarianism. The present trend where we see a movement toward increasing intolerance of alternative views is a sign that we could be traveling in the direction of totalitarianism.

Why is this happening and can the trend be reversed? There are indications that it begins in the school system. Teachers have mainly adopted a leftist ideology and some are aggressively imposing it on their students. Katie Hopkins (2017) has described this trend in simple terms by stating, "Schools are supposed to teach kids HOW to think, not WHAT to think. So why are so many liberal teachers bullying and brain-washing children with their own intolerant views?" In another article by Caroline Marcus (2016), she asks, "Why are Australian kids becoming dumb and dumber? Ask the teachers." The answer she gives is that "teachers are wasting time on ideological brainwashing, instead of focusing on literacy and numeracy." Students are being groomed in political correctness and gender and identity politics.

How can these trends be reversed?

In Chapter 7, we looked at innovations that could improve how school students are educated. One important omission in the chapter was how to educate students politically. Kate Habgood (2017) has some excellent suggestions how this might be done. She points to the political illiteracy of students in Australian schools. In the general population, there seems to be a pride in admitting to being apolitical. Habgood proposes some simple practical exercises to acquaint students with the concept of political ideology. She introduces the topic by asking students to draw the horseshoe mentioned earlier in the chapter. This then serves throughout the year as a template to help develop the students' own political opinions. Unlike the brainwashing described by Hopkins and Marcus, students are made aware of the good characteristics of both left and right ideology. They are then encouraged to form their own political views independently of outside influences. These views are flexible and can change over time.

How can the standard of debate be improved?

We saw (Chapters 2 and 6) that one of the dangers to science is authoritarianism. When dogmatic opinions are imposed and debate is silenced by one opponent, we trend toward totalitarianism. We are going to take three wise quotations on which to base a response to this trend,

The first is one by Evelyn Beatrice Hall: "I do not agree with what you have to say but I'll defend to the death your right to say it."[1] This statement epitomizes the ideal attitude to take in a debate. It illustrates the mutual respect that is needed to discuss an issue. There needs to be a total rejection of the position of some to try to impose their views, sometimes to the extent of using bullying, insults, and intimidation. Those who confront these tactics need to respond with firmness and courage. If one side of a political debate uses intolerant tactics, it is incumbent for those on the same side to pull them into line. In a successful democracy, it is essential for both sides of the political spectrum to be seen as respectful and authentic.

The second quotation is by Thomas Jefferson (but has also been attributed to John Philpot Curran): "The price of freedom is eternal vigilance."[2] Those who value freedom must be prepared to constantly defend it. They cannot afford to momentarily let their guard down.

The third quotation is by Edmund Burke: "The only thing necessary for the triumph of evil is for good men to do nothing."[3] When untruths are told, they need to be challenged, always of course, in a respectful way. If someone is treated unfairly, there are those among the observers who justify their inertia by declaring that it is not their problem. They need to recognize that it is their problem. Those who do nothing need to be cognizant of its effects.

Notes

1. http://en.wikiquote.wikiquote.org/wik/Evelyn_Beatrice_Hall, accessed July 2017.
2. www.quotationspage.com/quotes/4949html, accessed July 2017.
3. https://brainyquote.com/quotes/edmund_burke_377528.html, accessed July 2017.

References

Habgood, K. 2017. Political literacy in Australian schools. *Breaking Out* 1(1). breakingout.net.au/content/political-literacy-australian-schools.

Hopkins, K. 2017. Schools are supposed to teach kids HOW to think for themselves not WHAT to think. SO why are so many liberal teachers bullying and brainwashing children with their own intolerant view. *Daily Mail*, February 6. Accessed July 2017. www.dailymail.co.uk/news/article-4194048/KATIE-HOPKINS-liberal-brainwashing-schools.html.

Marcus, C. 2016. Why are Australians becoming dumb and dumber? Ask the teachers. *Daily Telegraph Melbourne*, December 20.

Wikipedia. n.d. Paradox of tolerance. https://en.wikipedia.org/wiki/Paradox_of_tolerance.

Index

A

Ad hominem, 106
Adler, 14, 16
Agenda 21 (sustainable development), 58–59
American Heart Association (AHA), 86
Anthropogenic global warming (AGW), 6, 8, 31
  greenhouse gases and, 98
  ideas relating to, 8
  opposing viewpoints on, 8
  political correctness and, 30
Authoritarianism, 20, 75–76, 120
Autoimmune disorders, 90

B

Bias
  elimination of, 110
  political, 60
  reinforcement, 4
  selection, 88, 122, 123
Blemished reputation, *see* Toxic workplace environment (CSIRO)
Bringing it together, *see* Interconnection of issues
Bullying, 42, 50
Bureaucracy, encroachment of, 89
Burke, Edmund, 130

C

Carbon dioxide dissolution, chemistry of, 84
Carbon tax, 98
Cardosa, Fernando Henrique, 21
Casey, R.G., 22–23
Celiac disease, gluten intolerance and, 90
Childhood education, role of wonder in, 101–102
Chilean military coup (1973), 55–56
Cholesterol, role of, 87
Churchill, Winston, 62, 113
Climate change, 79–80
  controversy, 29
  deniers, 98
CNN, 116
Commonwealth Scientific and Industrial Research Organisation (CSIRO), 22
  blemished reputation, *see* Toxic workplace environment (CSIRO)
  changes resulting from reviews of, 23–24
  establishment of, 23
  partnerships, 31
  short history after its formation, 23
  Victims of CSIRO blog site, 31–32
Consensus, 4, 9, 30
Control freak, 41
Controversies, *see* Current controversies, applying scientific thinking to
Coral bleaching effects, 83
Corporate psychopaths, 47
Creativity, 103
CRISPR (clustered regularly interspaced short palindromic repeats), 93
Critical thinking and scientific method, general introduction to, 4–11
  anthropogenic global warming, 8
  distinction between science and pseudoscience, 6–7
  faults in application of the scientific method, 8–9
  political debate, absence of scientific thinking in, 9–10

range of capacity for critical thinking, 5–6
status quo, 11
transition from dogmatic to critical thinking, 7
Cultural background, influence of, 107
Curran, John Philpot, 130
Current controversies, applying scientific thinking to, 79–95
acidification of the ocean, effects of, 83–84
autoimmune disorders, 90
biased thinking, 95
bureaucracy, encroachment of, 89
carbon dioxide dissolution, chemistry of, 84
celiac disease and gluten intolerance, 90
cholesterol, role of, 87
climate change, 79–80
CRISPR, 93
critical examination of controversy, 95
epidemiological research versus clinical studies, 88
food processors, problems for, 86
gene drives, 93
genetic engineering, 92
genetic engineering, risks, 94
genetic engineering, weighing the risks, 93–94
genetic manipulation, possibilities for, 92–93
gluten-free diets, downside of, 91–92
gluten-free foods, advent of, 90
gluten-free foods, explosion of, 91
gluten-free foods, need for, 90–91
Great Barrier Reef, 82–83
LDL and HDL cholesterol, 87–88
Medieval Warm Period, 81–82
need for critical examination of arguments, 84
observations consistent with previous ones, 80
saturated fat controversy, 85
saturated fat–heart disease hypothesis, 88
saturated fats, research on effects of, 85–86
selection bias, 88
status of the hypothesis, 89
superweeds, 94
terminator seeds, 94
trans fats, problem of, 86
trans fats, replacement of, 87
weakness of modeling, 80–81

D

Dangers to progress in science, 65–78
authoritarianism, 75–76
cases, 73–74
citations and impact factors for measuring merit, 68–69
earth as the center of the universe, 76
Enlightenment, 65
external and internal forces, effects of, 72
fraud in science, prevalence of, 74
Golden Age, 65
lack of citations to publications, 69
Lysenkoism, 76–77
maintaining the integrity of science, 75
managerialism, 77–78
misallocation of credit, 75
misconduct in science, 72
more subtle forms of unethical behavior, 73
negative flow-on effects from mediocre research, 70–71
not more scientists but more good scientists, 71
obsession with number of publications, 66–67
Piltdown Man, 72
poor-quality research, excess of, 69–70
Popper, dangers suggested by, 65–66
quantity and quality in scientific publications, 67–68
Darwin, Charles, 25, 26
Demagoguery, 64
Demarcation, problem of, 14
Democracies, pros and cons of, 113–114
Department of Agriculture (USDA), 86
DNA (deoxyribonucleic acid), 92
Dogmatic thinking
confirmations and, 7
scientific/critical versus, 110–111
transition from (to critical thinking), 7

E

Education
as means to raise standard of thinking, 123–124
system, changes needed in, 100–101
Einstein, Albert, 18, 24, 26, 103
Enlightenment, 65, 109, 120
Epidemiological research, clinical studies versus, 88

F

Fast Fourier transforms, 36
Fat–heart disease hypothesis, 88
Fearmongering, 64
Federal parliament, research scientists in, 98–100
Flip-flop, 97
Food processors, problems for, 86
Fortune-tellers, 17
Fox News TV network, 116
Franklin, Benjamin, 21
Fraud in science, prevalence of, 74
Freud, 14, 16
Future of critical thinking and scientific method, 127–130
- lessons from history, 127–128
- reversal of trends, 129
- standard of debate, improvement of, 129–130
- trends, 128–129

G

Galileo, 18
Gene drives, 93
Genetically modified (GM) crops, 94
Genetic engineering (GE), 92
Global warming deniers, 29
Gluten-free foods
- advent of, 90
- downside of, 91–92
- explosion of, 91
- need for, 90–91

Gluten intolerance, celiac disease and, 90
Golden Age, 65
Great Barrier Reef, 82–83, 122
Greenhouse gases, 1, 6, 82, 98

H

Hall, Evelyn Beatrice, 129
Hanson, Norwood, 14
Hare checklist, 44–45
High-density (HDL) lipoproteins cholesterol, 87–88
Hussein, Saddam, 54
Hypothesis
- description of, 7
- refuted, 18
- validity of, 121

Hypothetico-deductive method, 13

I

Ideology, prism of, 107
Imagination, 77, 118
Implementing scientific thinking and critical analysis, 97–107
- carbon tax, 98
- childhood education, role of wonder in, 101–102
- climate change deniers, 98
- creativity, 103
- cultural background, influence of, 107
- deflection of an issue, 105–106
- education system, changes needed in, 100–101
- federal parliament, research scientists in, 98–100
- flip-flop, 97
- how to introduce more scientific thinking, 100
- ideology, prism of, 107
- introducing critical analysis, 103–104
- paradox, 97
- putting labels on opponents, 106
- simple examples, 104–105
- skepticism, 99
- speculation, 106
- straw man argument, 106
- teaching with magic, 102–103

Informative content, 17
Interconnection of issues, 109–124
- authoritarianism, erosion of scientific progress by, 120
- bias, elimination of, 110
- cause–effect relationship, 121
- dangers to the progress of science, 117–118
- democracies, pros and cons of, 113–114
- determination that proposed new law is better than the law it replaces, 114
- diminished role of scientists, 111–112
- education as the means to raise the standard of thinking, 123–124
- electorate, capacity for critical thinking, 114–116
- Enlightenment, 109
- flaws in the application of thinking to some current issues, 120–123
- how managerialism erodes the standard of scientific research, 118
- how scientific organizations can be infiltrated by destroyers, 113

lack of imagination or real interest, 118
media, 116
opinion writers, 117
psychopaths, dangers posed by, 124
public opinion, influence of media on, 116–117
reasons that science may not be advancing as it should, 119–120
science progression, 118–119
scientific/critical thinking, 109–110
scientific/critical versus the dogmatic approach, 110–111
scientific excellence, effects on, 112
spirit of inquiry, 124
Intergovernmental Panel on Climate Change (IPCC), 29, 98
Internet, development of, 66
Iraq, invasion of (2003), 54

J

Jefferson, Thomas, 130

K

Kepler, 18
Kordig, Carl, 14
Kuhn, Thomas, 14

L

Langmuir, Irving, 25, 26
Legal argument, 46
Low-density (LDL) lipoprotein cholesterol, 87–88
Lysenkoism, 20, 76–77

M

Magic, teaching with, 102–103
Magnetic resonance imaging (MRI), 47
Managerialism, 77–78, 118
Marx, 14
Matrix system, 31
Media, influence of (on public thinking), 53–64, 116–117
Agenda 21 (sustainable development), 58–59
British immigration since early, 1980s, 59–60
Chilean military coup (1973), 55–56
contrast between submissions to scientific journals and to mainstream press, 57–58
deciding what is the truth, 56–57
demagoguery, 64
fearmongering, 64
how a democratic system can break down, 62
how to think for yourself, 63–64
invasion of Iraq (2003), 54
issues that seem to be ignored by the mainstream media, 58
myth of the lemmings, 53–54
opinion writers, 60–61
social media, rise of, 63
stereotypes, 64
Medieval Warm Period (MWP), 81–82
Merkel, Angela, 21
Mobbing, 42
Modeling
illusion of, 4
weakness of, 80
MSNBC, 116

N

Narcissist, 41–42
National Council of Teachers of English (NCTE), 109
Newspapers, 60
Newton, 18
NIH, *see* U.S. National Institutes of Health

O

Ocean acidification, effects of, 83–84
Olive oil, 87
Opinion writers, 60–61, 117

P

Paradigm shifts, 14
Pasteur, Louis, 25, 26
Phytoliths, 82
Piltdown Man, 72
Pinochet, Augusto, 55
Plato, 5
Political bias, 60
Political candidates, 62
Political correctness, 30, 34, 99

Political debate, absence of scientific thinking in, 9–10
Political illiteracy of students, 129
Political thinking, distinction between scientific thinking and, 97
Politics, why more scientists do not enter, 34–35
Popper, Karl, 3, 14, 6, 118, 128
Positron emission tomography (PET), 47
Probability, 17
Pseudoscience, distinction between science and, 6–7
Psychopath(s), 42–43
  characteristics of, 43–44, 44–45
  corporate, 47
  dangers posed by, 124
  description of, 43
  identification of, 44
  in science, 46–47
Psychopathic behavior
  coping with, 48
  example, 45
  origin of, 47
Public opinion, influence of media on, 116–117

R

Refutability, 3, 16
Relativity, theory of (Einstein), 15, 24
Research organizations, lack of scientific input in (impact of), 21–32
  Australian ministers in charge of science portfolio, 21–22
  balanced debate, 30
  changes resulting from reviews of CSIRO, 23–24
  climate change controversy, 29
  comparison of the two approaches, 26–27
  effect of changes on workplace environment, 28
  errors in application of science, 29–30
  expansion of the managerial approach, 30–31
  Honorable R.G. Casey, contribution of, 22–23
  how CSIRO has fared, 27
  how managerial control purports to drive science, 25–26
  how scientific knowledge is acquired, 24–25
  lack of scientific thinking in the debate, 29
  matrix system, 31
  present working environment, 31
  requirement to obtain a proportion of research funding, 27–28
  short history of CSIRO after its formation, 23
  Victims of CSIRO blog site, 31–32

S

Saturated fat controversy, 85
Scientific method, 13–20
  authoritarianism and, 20
  comparisons of different theories, 15
  conjectures and refutations, 18
  criteria for evaluating a theory, 18
  dangers to progress, 19–20
  guidelines for evaluating theories, 16
  hypothetico-deductive method, 13
  lack of imagination, 19
  Lysenkoism, 20
  need for science to grow (or to progress), 18–19
  notable theories of the early twentieth century, 14–15
  paradigm shifts, 14
  probability and informative content, 17
  problem of demarcation, 14
  refutability as the criterion for demarcation, 16–17
  requirements for progress of science, 19
Scientific method, common misunderstandings of, 1–4
  consensus is not a criterion for the validity of a theory, 4
  illusion of modeling, 4
  refutability as a criterion for evaluating a scientific theory, 3
  scientific theory can never be proven beyond doubt, 2
  separate observations consistent with theory do not correspond to additional confirmation, 3
  topic of debate must be defined unambiguously, 1
  true scientist must be detached, 2
Selection bias, 88, 122, 123
Serial bully, 42
Skepticism, 99
Social media, rise of, 62, 63

Southern South America (SSA), 82
Speculation, 106
Spirit of inquiry, 124
Status quo, 11, 63
Stereotypes, 64
Straw man argument, 106
Superweeds, 94
Szent-Györgyi, Albert, 25, 40

T

Teaching with magic, 102–103
Terminator seeds, 94
Thatcher, Margaret, 21
Totalitarianism, 179
Toulmin, Stephen, 14
Toxic workplace environment (CSIRO), 33–50
 career in science, 33–34
 control freak, 41
 corporate psychopaths, 47
 decisions made by politicians, 39
 effect of organizational changes on research, 38
 effect of requirement to procure funding, 34
 Hare checklist, 44–45
 how important scientific discoveries are made, 38
 how it happened, 49–50
 how psychopaths attain high positions in corporations, 48
 how science of excellence can still be achieved, 36–38
 how toxic workplace environments arise, 40
 legal argument, 46
 major turning point, 37
 making decisions without the basic knowledge, 37
 narcissist, 41–42
 personality differences in scientists, 35–36
 psychopath, 42–43
 psychopath, characteristics of, 43–44, 44–45
 psychopath, description of, 43
 psychopath, identification of, 44
 psychopathic behavior, coping with, 48
 psychopathic behavior, example, 45
 psychopathic behavior, origin of, 47
 psychopaths in science, 46–47
 scientific leadership, 39
 serial bully, 42
 why more scientists do not enter politics, 34–35
 why toxic work environments form in science organizations, 40
 workplace bullying, 50
Trans fats, 86, 87
True science, 5

U

USDA, *see* Department of Agriculture
U.S. National Institutes of Health (NIH), 25, 86

V

Victims of CSIRO blog site, 31–32, 49
Voting, 115

W

Western world financial crisis, 47
Wireless local area networking (WLAN), 36
Wireless networking technology (Wi-Fi), 36
Workplace environment, toxic, *see* Toxic workplace environment (CSIRO)